Spectroscopy in the Biomedical Sciences

Editor

R. Michael Gendreau, M.D., Ph.D.
Manager
Biomedical Instrumentation Office
Biotechnology Department
Battelle Memorial Institute
Columbus, Ohio

CRC Press
Taylor & Francis Group
Boca Raton London New York

CRC Press is an imprint of the
Taylor & Francis Group, an **informa** business

First published 1986 by CRC Press
Taylor & Francis Group
6000 Broken Sound Parkway NW, Suite 300
Boca Raton, FL 33487-2742

Reissued 2018 by CRC Press

Library of Congress Cataloging-in-Publication Data

Spectroscopy in the biomedical sciences.

   Bibliography: p.
   Includes index.
   1. Spectrum analysis.   2. Biomolecules—Analysis.
I. Gendreau, R. Michael.
QP519.9.S6S64   1986   574.19'285   86-6145
ISBN 0-8493-5740-3

A Library of Congress record exists under LC control number: 86006145

Publisher's Note
The publisher has gone to great lengths to ensure the quality of this reprint but points out that some imperfections in the original copies may be apparent.

Disclaimer
The publisher has made every effort to trace copyright holders and welcomes correspondence from those they have been unable to contact.

ISBN 13: 978-1-315-89773-8 (hbk)
ISBN 13: 978-1-351-07683-8 (ebk)

Visit the Taylor & Francis Web site at http://www.taylorandfrancis.com and the
CRC Press Web site at http://www.crcpress.com

PREFACE

The intent of this collection of manuscripts is to provide general descriptions of analytical techniques which we believe to be useful to the biological scientist, and to provide examples of the utility of each technique. The contributing authors were asked to focus on examples in which their particular technique has proven particularly useful in studies of biological systems. Many commonly used analytical techniques such as NMR and ESR are not included in this work; as editor, I chose to focus on techniques which perhaps have not received as much coverage as in the recent literature.

The authors who have contributed to this work and I believe that all of the analytical tools covered in this volume have great utility, and the scientific community can expect to see increasing usage of most if not all of these techniques over the next few years.

R. M. Gendreau
Columbus, Ohio
November, 1984

THE EDITOR

R. Michael Gendreau, M.D., Ph.D. is currently serving as Manager of the Biomedical Instrumentation Program Office at Battelle Columbus Division, Columbus, Ohio. He is also Principal Investigator of Battelle's National Center for Biomedical Infrared Spectroscopy, a national facility funded by the National Institutes of Health, Division of Research Resources.

Dr. Gendreau obtained his M.D. and Ph.D. from the Ohio State University, Columbus, Ohio. He is a member of numerous professional societies, including the American Medical Association, Ohio State Medical Association, Society for Biomaterials, Society for Applied Spectroscopy, and the American Chemical Society. He served as plenary lecturer at the 1985 International Conference on Fourier and Computerized Infrared Spectroscopy. Dr. Gendreau also teaches a series of short courses on biomedical applications of FT-IR spectroscopy at various locations around North America. He has authored or co-authored over 20 publications and one book covering spectroscopy, instrumentation, obstetrics, pharmacology, and biomaterials. He has presented numerous research papers at international meetings, and has been selected for many awards, such as the Max Weiss Award for outstanding research by a graduate medical student in 1981.

Dr. Gendreau's current research efforts are in the field of biomedical research, specializing in areas such as medical instrumentation development, bioengineering and diagnostics development, medical device and product development, health care delivery assessment, and biomedical spectroscopy.

CONTRIBUTORS

Ian M. Armitage
Department of Molecular Biophysics
  and Biochemistry
Yale University
New Haven, Connecticut

David G. Cameron, Ph.D.
Research Fellow
Molecular and Surface Spectroscopy
The Standard Oil Company
Cleveland, Ohio

Seth A. Darst, B.S.Ch.E.
Research Assistant
Department of Chemical Engineering
Stanford University
Stanford, California

Richard A. Dluhy, Ph.D.
Research Scientist
Biotechnology Department
Battelle Memorial Institute
Columbus, Ohio

R. Michael Gendreau, M.D., Ph.D.
Manager
Biomedical Instrumentation Office
Battelle Memorial Institute
Columbus, Ohio

Joseph L. Giegel, Ph.D.
President
Gulfstream Diagnostics Corporation
Miami, Florida

Alan G. Marshall, Ph.D.
Professor
Departments of Chemistry and
  Biochemistry
The Ohio State University
Columbus, Ohio

Brien J. McElroy, M.S.
ESCA Laboratory Coordinator
Department of Chemical Engineering
University of Washington
Seattle, Washington

Richard Mendelsohn, Ph.D.
Professor
Department of Chemistry
Rutgers University
Newark, New Jersey

Laurence A. Nafie, Ph.D.
Professor
Department of Chemistry
Syracuse University
Syracuse, New York

Buddy D. Ratner, Ph.D.
Associate Professor
Department of Chemical Engineering
University of Washington
Seattle, Washington

Channing R. Robertson, Ph.D.
Professor
Department of Chemical Engineering
Stanford University
Stanford, California

TABLE OF CONTENTS

Chapter 1

BIOMEDICAL FOURIER TRANSFORM INFRARED SPECTROSCOPY: EQUIPMENT AND EXPERIMENTAL CONSIDERATIONS

R. Michael Gendreau

TABLE OF CONTENTS

I. INTRODUCTION

A. Overview

The application of infrared spectroscopy (IR) to studies of chemical systems is not new — in fact, IR has been widely used in organic and analytical chemistry for many years. Application of IR techniques to biological systems has been less common and in general has met with less success, and as a result IR has not been regarded as a very useful tool for studies of aqueous biological systems. The development and evolution of Fourier transform infrared spectrometer (FT-IR) systems has changed this, and FT-IR has now become a valuable tool for studies of aqueous biological systems. Our group at Battelle has been exploring biological applications of FT-IR since the mid 1970s, and the technique has proven to have a number of worthwhile biological applications. The improvement in instrument performance since the mid 1970s has been truly remarkable, and is responsible in large part for the current renewed interest in biological IR studies. The National Institutes of Health has established the National Center for Biomedical Infrared Spectroscopy at Battelle, and with this Center in operation, we expect to see an increased interest in biological FT-IR and an increase in the number of users of biological FT-IR.

This chapter is intended to present the basic principles of FT-IR, and the special considerations for investigators interested specifically in biological FT-IR. I have in general followed the notation of Griffiths when mathematically treating interferograms and spectra, and the reader is referred to this work for a more thorough discussion of FT-IR principles.[1]

It has been known for many years that most organic and biological molecules absorb long-wavelength light in the infrared region, at frequencies and with extinction coefficients dependent on the exact molecular nature of the system under study. This property has been exploited to provide an analytical technique which may be used to characterize materials interacting with infrared light, and this characterization principle has led to the "fingerprint" concept. That is, each molecule gives a distinctive fingerprint pattern of infrared light absorption, and this pattern can be used for identification of unknowns. Because the light absorption patterns are functionally group-specific, this technique is well suited for detecting changes in molecules such as those due to addition or loss of functional groups, changes in hydrogen bonding, and so on.

IR as an analytical tool first emerged in the 1940s when the first commercial dispersive infrared spectrometers became widely available. Their development was spurred by World War II, and over the next 20 years infrared spectrometers slowly evolved in an incremental fashion as improvements in grating and detector technology took place. However, the evolution of IR was much slower and less impressive than the evolution of other analytical techniques such as mass spectroscopy and nuclear magnetic resonance (NMR) spectroscopy. Thus, by the late 1960s, many scientists were of the opinion that IR would never again be a first-line analytical tool, and that NMR and mass spectroscopy were of much greater utility, especially in the biological area. Among vibrational spectrosopists, Raman spectroscopy was felt to be the preferred biological technique. However, during the late 1960s, the advent of commercial FT-IR took place, thus beginning a slow revolution in the field of IR. From the early 1970s through today, there has been a continuing dramatic improvement in hardware and software related to FT-IR spectroscopy. These developments have rekindled an interest in IR as a general science, and in the biomedical applications of IR in particular.

B. Fourier Transform Spectroscopy

Traditional grating-based (dispersive) IR was generally plagued with an undistinguished sensitivity, a relatively low data acquisition rate, some significant problems

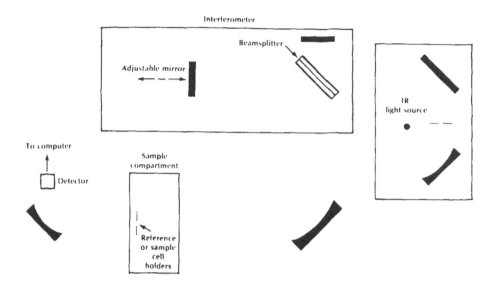

FIGURE 1. Schematic diagram of a conventional Michelson interferometer. Also illustrated are the source compartment, sample compartment, and detector compartment.

with long-term reproducibility, and a general lack of molecular specificity. The general lack of specificity results from the fact that every organic or biological molecule in a system causes infrared light to be absorbed, and interpretation of the results is often quite difficult because of band overlaps and band-assignment problems. So while the infrared region is information rich, in many ways there is too much information, and interpretation is often complex.

As an alternative to the dispersive approach, there is the multiplex (multiple channel or Fourier transform) approach. The Fourier transform approach involves encoding wavelength-related information in the source output, and the full modulated source output is passed through a sample continuously. The signal detected represents all wavelengths of light from the broad band mid-infrared source, and a mathematical sorting process is undertaken to decipher the interactions which take place when infrared light is absorbed by a molecular system. Figure 1 shows a diagram of a conventional Michelson interferometer design for a mid-infrared spectrometer. The Michelson interferometer typically consists of a 45° beam splitter, which ideally reflects 50% of the incident light to a fixed mirror, and transmits 50% to a movable mirror. As the movable mirror is scanned, the path difference between the two mirrors and the beam splitter change continuously. The two beams of light are reflected by the mirrors back to the beam splitter and recombined, but the beams are now out of phase with each other. Since the length of the optical path for the moving mirror arm of the interferometer changes, the light in this arm is continuously changing phase relative to the fixed arm, and a pattern of constructive and destructive interference takes place. If only

monochromatic light is passed through a Michelson interferometer, the result is a cosine wave. Initially, the path difference in the interferometer is equal, and a maximal signal is measured. As the movable mirror begins to scan, the light in the two arms of the interferometer begins to undergo destructive interference. Eventually the light will become fully out of phase and complete destructive interference will result. As the mirror continues to scan, the light cycles back in phase once again and another maximum is reached. The frequency of this cosine wave is determined by the wavelength of the light used, and the initial amplitude of the cosine wave is determined by the intensity of the light of the wavelength giving rise to that cosine wave. If an infinitely narrow band of light is put in an interferometer (a laser, for example), a pure cosine wave will

Table 1

INFORMATION IN FOURIER SPACE AND SPECTRAL
SPACE

Spectral Space	Fourier Space
Band frequency or wavelength	Cosine frequency
Band intensity	Initial cosine amplitude
Band half-width	Rate of cosine amplitude decay
Band profile (Lorentzian, Gaussian, etc.)	Cosine decay profile

result. However, as bands become broader (with a finite half width), the cosine waves become exponentially damped. It can be shown mathematically that the Fourier transform of a decaying exponential function is a Lorentzian function; therefore, the cosine wave caused by a band of Lorentzian profile will have an exponential decay in intensity as the mirror is scanned. The rate of decay is dictated by the half width of the Lorentzian band in question. Thus, a Lorentzian infrared band will have several distinguishing features in the interferogram (or Fourier space), and these are summarized in Table 1.

The mathematical process of Fourier transformation allows conversion from Fourier space to spectral space and vice versa. The information content in either domain is equivalent, and conversion back and forth between domains is quite straightforward.

C. Instrumental Resolution

Resolution of two closely spaced bands by interferometry requires that the two cosine waves which arise from these two bands be differentiated. To differentiate two closely related cosine waves, the mirror must be scanned far enough so that the two cosine waves have the opportunity to cycle out of phase with each other. Thus, resolution of two bands occurs when the mirror has moved far enough for the two cosine waves to become out of phase one cycle. This is based on the path difference induced in the interferometer, which is directly related to the distance the movable mirror is scanned. The relationship is that the instrumental resolution (in wavenumbers) is improved proportionally with increasing mirror stroke distance by the following formula:

$$\Delta v = \frac{1}{2\Delta x} \tag{1}$$

where Δv is the resolution in wavenumbers and Δx is the mirror stroke in centimeters.

D. The Advantages of Fourier Transform Infrared Spectroscopy

The tremendous progress that has been made in the last 10 years in IR is primarily related to advances made in the technology of FT-IR spectrometers. FT-IR instruments, as compared to dispersive instruments, have several significant advantages which are described below:

1. The multiplex advantage. A large number of resolution elements or wavelengths can be monitored simultaneously in FT-IR. This results in a signal-to-noise ratio (SNR) per unit time improvement proportional to the square root of the number of resolution elements being monitored.
2. The throughput advantage. Because interferometry does not require use of a slit or other restricting device, especially at low resolution, the total source output can be passed through the sample continuously. This results in a fairly substantial gain in energy at the detector (throughput), which translates to higher signals, and generally, improved SNRs.

3. The speed advantage. Because it is practical to move the mirror in interferometry short distances quite rapidly, and because of the SNR improvements due to the multiplex and the throughput advantages, it is quite realistic to obtain useful infrared spectra on a millisecond time scale. This obviously has important ramifications for studying real-time biological systems and reactions.

4. Frequency precision. Although not often mentioned as a typical FT-IR advantage, the presence of the helium-neon laser to reference the mirror position provides for great frequency precision in the final spectrum obtained after Fourier transformation. When performing interferometry, the factor that determines the precision with which the position of an infrared band can be determined is the precision with which the scanning mirror position is known. By using a laser as a reference, mirror position is known with extremely high precision. As a result, the Fourier transformed spectrum has the same level of precision in band position. It has been estimated that FT-IRs are precise to ± 0.002 cm^{-1} when operating at 2 cm^{-1} resolution. This translates directly to an increased ability to perform spectral subtraction and other related mathematical processes, and is perhaps the major reason why computerized grating infrared systems do not perform subtractions as well as FT-IR systems.

In summary, the advantages of Fourier transform infrared instrumentation, starting with the multiplex and throughput advantages, and then considering the continuing hardware improvements in terms of detector performance and instrument stability, have greatly increased the applicability of IR to biological systems. Over the last decade, we have witnessed at least an order of magnitude improvement in instrumental sensitivity, and a corresponding decrease in the minimum time required to obtain a complete infrared spectrum. Today it is realistic to obtain infrared spectra from biological systems on a 100-msec time scale, and to then carry out a number of mathematical manipulations on those data. The SNRs that can be obtained today are truly outstanding, and they translate directly to an ability to probe much more subtle changes in molecular systems, and an ability to decipher information from very complex systems which would otherwise be unapproachable by a relatively nonspecific technique such as IR.

II. THEORY

A. The Michelson Interferometer

Most commercial FT-IR spectrometers, with the exception of the Analect design, use a Michelson interferometer as the basis for the infrared analysis. The Michelson interferometer shown in Figure 1 is a two-beam design in which infrared radiation is split into two paths and the relative path lengths are varied with respect to one another.

Michelson first described the interference effects which result from use of this design interferometer in the 1880s. The realization by Rayleigh in 1892 that the information collected using a Michelson interferometer (the interferogram) is related to a conventional spectrum through the Fourier transform, i.e., the fact that they are transform pairs, paved the way for the application of interferometry to IR. Historically, infrared interferometry was developed for high-resolution physics and astronomy applications, but more recently, chemical applications have become predominant. As the technology has moved from physics to chemistry and now to biology and medicine, the requirements in terms of instrumental performance have changed dramatically. Later in this chapter the specific instrumental features which are desirable for biomedical interferometry are discussed.

As has been mentioned, the interferometer encodes spectral information based on an induced path difference created by moving a mirror. One of the primary differences between various vendors of FT-IR systems is the transport mechanism for the movable mirror. Perhaps the most common design uses an air bearing developed by Curbello of Digilab, which has the mirror moving on a cushion of air. Another design uses a mechanical bearing, and oil bearings have been used as well. Each of the scanning approaches has advantages and disadvantages, but all are designed to make the movement of the mirror as smooth, tilt-free, and vibration-free as possible. The moving mirror must stay perpendicular to the light striking it to a tolerance of well under a wavelength of light. Any vibration, tilt, or speed change during the sweep of the mirror will cause instability, systematic noise, "glitches" or spikes, or some combination of these. Therefore, the drive mechanism of the interferometer is an essential component in the overall functioning of the interferometer. Several of the vendors have optical or mechanical designs to minimize the sensitivity of their interferometer to mirror tilt and vibration.

As an example of how a change in path difference between two arms of a Michelson interferometer can be used as a spectral coding technique in a spectrometer, consider what occurs when the beams are recombined at the beam splitter prior to detection. As can be seen in Figure 1, the light beams are recombined as they pass through the beam splitter, and one half of the energy is then directed through the sample compartment and to the detector. This recombined beam will undergo totally constructive interference if the path difference or retardation, Δ, is an integral number of wavelengths of the analyzing wavelength, i.e., $\Delta = n\lambda$ where n is an integer. (For the sake of simplicity, we will assume for the moment that we are dealing with a single monochromatic wavelength.) It follows that totally destructive interference occurs at $\Delta = (n + 1/2)\lambda$. In the event of no outside distrubances, a monochromatic ray of light being passed through a Michelson interferometer will give a cosine wave of continuously varying amplitude as the mirror is scanned. The frequency of this cosine wave is a function of the speed of mirror movement and wavelength of light selected. The equation to express the amount of energy being passed by the interferometer at any mirror position is given by the following equation. For monochromatic light of frequency ν:

$$I(\delta) = 0.5I(\nu)[1 + \cos(2\pi\nu\delta)] \tag{2}$$

where $I(\delta)$ = the total interferogram intensity at mirror position δ and $I(\nu)$ = the total energy available (total intensity) at frequency ν. Note that as δ is increased (the mirror is scanned), the interferogram intensity varies as a cosine function.

For polychromatic light, we integrate and correct for instrumental factors:

$$I(\delta) = \int_0^\infty B(\nu) \cos(2\pi\nu\delta)d\nu \tag{3}$$

where $B(\nu) = 0.51(\nu)C(\nu)$ and $C(\nu)$ = an instrumental correction factor.

The other half of the Fourier transform pair which evaluates $B(\nu)$ as a function of mirror position is

$$B(\nu) = \int_{-\infty}^\infty I(\delta) \cos(2\pi\nu\delta)d\delta \tag{4}$$

This equation is then evaluated to produce the infrared spectrum, $B(\nu)$ of light intensity vs. frequency, starting from a data set $I(\delta)$, or light intensity vs. mirror position. In practice, optical and electronic imperfections cause phase distortions in the optical beam, and a phase term is added to the final equation:

$$B(\nu) = \int_{-\infty}^{\infty} I(\delta) \cos(2\pi\nu\delta - \theta_\nu)d\delta \tag{5}$$

where $\theta\nu$ is a correction term known as the phase angle.

Expressed another way, the signal measured by the detector in a Fourier transform spectrometer is described as an interferogram, a summation of an infinite number of cosine waves, all of which add constructively or destructively, depending on their individual phases at each position of the scanning mirror. The addition of a sample to the beam causes these relationships to change due to absorption of light by the sample which selectively removes certain frequencies. The Fourier transformation, according to Equation 5 above, can be thought of as an unshuffling where each of the cosine waves is related back to energy transmitted by the sample as a function of frequency.

B. Fourier Data Processing

As Fourier transformation is an integral part of obtaining an FT-IR spectrum, we should not overlook the importance of this step of the operation. The current fast Fourier transform algorithm was described by Cooley and Tukey in the mid 1960s, and this algorithm has greatly decreased the time required to calculate a Fourier transform by digital computer.

In addition to the Fourier transformation, several other data-processing steps are generally carried out:

1. Apodization. The fact that the mirror is not scanned an infinite distance implies that at some point data collection is abruptly stopped. This truncation of data collection causes a line shape function to be applied to the final spectrum, i.e., the shapes of the experimentally measured bands are altered by instrumental factors. Various mathematical functions are usually applied to the raw interferogram to "apodize" (to make without feet) the data. The apodization function causes the line shape function to be changed, and usually these functions minimize negative side lobes which are a problem if no apodization function is used.
2. Phase correction. The phase lag term in the Fourier transform due to optical and electronic factors is evaluated by a phase correction algorithm, which corrects the data set for the lag.
3. Zero filling. By simply adding zeros to the end of the interferogram, a spectrum with twice the sampled number of data points is obtained. Zero filling in Fourier space is essentially the same as interpolation in spectral space, and one level of zero filling improves the digital definition of the measured spectrum and should generally be used.

In addition to the advent of the FFT algorithm, the vast improvement in mini- and microcomputer performance over the last decade and the appearance of parallel (array) processors have greatly improved the speed and convenience with which Fourier transforms can be performed. In fact, the total time from data acquisition to data transformation and computation is often only slightly longer than data acquisition time per se. Thus, in contrast to the early days of FT-IR in which the time considerations of performing an FFT greatly limited the usability of the technique, compute time is no longer a major consideration in most experimental work. The instrumentation we use in our laboratory, with a dedicated minicomputer, an array processor which is used as a coprocessor, and high-speed Winchester disk drives, will perform a 4000-point Fourier transform in times approaching 100 msec or less. We have found that these speeds are desirable because many of our real-time applications generate tremendous amounts of data, and the ability of the data system to keep up with data acquisition in real time greatly simplifies post processing. Also, this speed provides feedback in real time about the progress of the rather complex experiments being carried out.

C. Trading Rules in FT-IR

With respect to biomedical applications of FT-IR, it can be said that resolution, contrary to many popular myths, is generally not an important consideration. Many commercial FT-IR systems achieve resolutions in excess of 0.5 cm^{-1}. Over the conventional mid-infrared range of data acquired by an FT-IR, this translates to roughly 32,000 data points acquired. Several undesirable things happen when this much data are obtained. Number one, it takes a large amount of disk space to store and manipulate. It also greatly slows down the computer (all the computations are lengthened by the number of data points which must be manipulated), but worst of all, running at unnecessarily high resolutions degrades the signal-to-noise obtained per mirror scan (time).

Consider what resolution is required for conventional, aqueous-based biomedical FT-IR measurements. Most biological species discussed in this book are macromolecules that have very complex infrared vibrations. As a result, the infrared absorptions tend to be quite broad due to smearing of the energy levels in the macromolecules and the mixing of vibrational modes.

Most protein bands range from 15 to 50 cm^{-1} in half-width at half-height (HWHH), and it is easy to demonstrate that resolutions greater than (numerically smaller than) half the HWHH of a band are not required to accurately describe the sample. Furthermore, most complex macromolecules have bands which are inherently overlapped; that is, the infrared absorptions physically overlap, and no degree of instrumental resolution can separate these bands. What, then, are the consequences of running at too great a resolution? This brings us to a description of the trading rules.

First, the signal-to-noise ratio per scan in a commercial FT-IR is a direct function of the resolution. That is, if the resolution desired is doubled, SNR obtained per scan will only be half as great.

Second, the time to obtain a scan is a direct function of the resolution. To double the measurement resolution, the mirror travel in the Michelson interferometer must be doubled. Since mirror velocity is constant, this requires an effective doubling in the measurement time per scan.

Third, at resolutions greater than roughly 2 wavenumbers, the source must be apertured to more fully columnate the infrared radiation being passed through the interferometer. This factor decreases the SNR obtained per unit measurement time, but it is variable, depending on the interferometer design and the degree of aperturing required to columnate the beam sufficiently. If the problem with required aperturing is ignored, doubling the instrumental resolution of the spectrum requires a minimum of a factor of 4 increase in measurement time to maintain the SNR — twice the number of scans, at twice the time per scan. Additionally, computational times, display times, and media storage requirements are all increased proportionally with increasing resolution. Thus, always run at the lowest resolution that is sufficient to accurately measure the infrared bands of interest. We have found in general, and this has been corroborated by other workers in the field, that 8 wavenumbers is generally very adequate resolution for macromolecular studies by FT-IR. In fact, as will be discussed later in this chapter, the SNR gained by running at the lower resolution allows more mathematical manipulations such as spectral subtractions and Fourier deconvolutions. We ultimately learn more about the sample by obtaining an improved SNR than we would by increasing instrumental resolution which does not provide additional information if the bands of interest are broad.

D. Detectors for Mid-Infrared FT-IR

The thermocouple-type infrared detector which has been used for many years in grating instruments is generally unsatisfactory for FT-IR applications because is is too slow to respond to thermal changes. The high-modulation frequencies (5 to 50 kHz) imposed by the movement of the scanning mirror of the FT-IR require a detector with a relatively rapid response. Two classes of detectors are in general use at the present time. The more-or-less standard detector which is sold with new commercial FT-IR systems is the triglycine sulfate (TGS) pyroelectric bolometer. This detector has the advantages of being inexpensive, of having a relatively rapid response, of being highly linear over a wide range of energy inputs, and of being operable at room temperature. It suffers from the disadvantage of being relatively insensitive compared to the thermocouple or the liquid-nitrogen-cooled class of detectors.

Nonetheless, this is a satisfactory detector for applications where the infrared transmission of the sample and/or the sampling accessory is very high. That is, in cases where energy throughput is not a limitation, a TGS detector may perform well. However, in the majority of samples of interest to biologists, including those in water, the higher sensitivity detector is required. The detector of choice in this case is generally the mercury-cadmium-telluride (MCT) liquid nitrogen cooled (77 K) photoconductive detector. The MCT class of detectors is much more sensitive than TGS detectors, often by a factor of nearly 100. Relating this back to the trading rules, this means that a high sensitivity, narrow-range MCT detector under some conditions can obtain a spectrum with the same SNR as that collected with a TGS in 1% of the measurement time. The MCT detector suffers from a lesser linearity, a greater cost, the need to operate at liquid-nitrogen temperatures, and more stringent electronics requirements in the preamplifier circuitry. Nonetheless, for those serious about biomedical applications, the MCT detector is a must. Especially in cases where attenuated total reflection optics are being used and a considerable amount of energy is lost due to the sampling process itself, the sensitivity of an MCT is essential.

Having decided to use a mercury-cadmium-telluride detector, there is still a choice of what "flavor" of MCT to choose. The MCT photoconductor chip can be specified as broad-range, mid-range, or narrow-range, defined in terms of the frequency response. The range determines the wavenumber or wavelength at which the element has its cut-off point. The longer the wavelength of light an MCT detector can respond to, the less energetic are the photons being detected, and thus, the more closely spaced the quantum levels in the MCT lattice must be. MCTs of this configuration have a much lower efficiency, and often times will yield a sensitivity only three- to fivefold improved over that obtained with a TGS. These so-called broad-band detectors often have cutoffs at 450 wavenumbers or lower, and are very useful for transmission experiments in aqueous systems. These systems benefit from a detector with a wide frequency response range, and do not require the sensitivity that the narrower range MCTs provide. The other extreme is the narrow-range MCT with very high sensitivity, a high-frequency cutoff, and a relatively small linear operating range. These detectors typically cut off around 800 wave numbers. Narrow range detectors are unable to detect lower-energy photons, and this detector will often have sensitivities as high as two orders of magnitude greater than those obtained with TGS room temperature detectors. Another additional benefit of the MCT class of detectors is the fact that one of the sources of inherent detector noise, the so-called 1/F noise, is reduced by modulating the radiation at higher frequencies. This has allowed the design of interferometers that scan more quickly, and the performance of the MCT detector, both in terms of D* (a measure of detector sensitivity) and 1/F performance, can be shown to improve somewhat as one goes to the higher scan rates. In our laboratory, we find that by increasing the scan speed of the mirror by a factor of 4, and thus decreasing the measurement time by a factor of 4, we obtain essentially the same SNR per scan with our MCT detector; that is, almost all of the SNR degradation that one would expect due to shorter measurement time is offset by the improvement in performance of the MCT detector and associated electronics at that scan rate. Another advantage is the fact that the faster one scans, the quicker one obtains the desired interferogram. In cases where adequate SNR can be obtained in a single sweep of the mirror, improvements in the scan speed of the spectrometer allow shorter measurement times, and allow experiments such as kinetic determinations of relatively fast reactions.

The latest generation of commercial FT-IR systems, including the Nicolet 60SX, the Digilab FTS 15/90, and Mattson Sirius, have provided spectrometers which are capable of collecting in excess of ten spectra per second. With this time resolution of approximately 100 msec, a significant number of biomedically interesting reactions and events can be followed in a more-or-less continuous fashion. The primary prerequisite of this experiment is that one be able to obtain a useful SNR in the measurement time of 100 msec. Commercial instrumentation is beginning to approach the point where this is in fact the case. However, when 40,000 to 50,000 data points per second are collected, the data processing, data transfer, and data storage problems become as significant as the problem of obtaining the spectrum with the SNR required.

E. Dynamic Range

Absorption spectroscopy in FT-IR puts very stringent constraints on the data acquisition system and particularly on the detector, preamplifier, amplifier, and analog-to-digital (A/D) conversion circuitry. In fact, the dynamic range provided by current state-of-the-art A/D converters is one of the fundamental limitations of FT-IR. The SNR required in an interferogram to obtain a corresponding signal-to-noise in the transformed spectrum is described by the following equation:

$$SNR_I = \sqrt{m} \, SNR_s \tag{6}$$

where m = the number of resolution elements collected, SNR_I = the interferogram SNR, and SNR_s = the desired spectral SNR.

If an infrared spectrum is collected at 8 wavenumber resolution, this gives roughly 2000 resolution elements (spectra obtained from 0 to 7900 wave numbers before transformation). Even if a technique called undersampling is used, which requires optical filtering and restricts the maximum measurement to 3900 wave numbers, one is still dealing with 1000 resolution elements. Consider a sample of an aqueous protein. In the experimental apparatus used in our laboratory, the absorbance due to the water band is typically in the range of 0.8 Absorbance Units (AU), and it is often desirable to measure bands as small as 0.05 mAU with a 4 to 5 SNR. This requires a SNR in the parent spectrum (in which the strongest signal is the water absorbance) of 50,000:1 before water subtraction. If this is related back to the signal-to-noise required in the interferogam, the required SNR is $50,000 \times \sqrt{1000} \simeq 1,500,000$ SNR. The full dynamic range of a 15-bit ADC is 32,000:1, which indicates that if the detector is capable of providing a 15-bit dynamic range signal, with noise being just the last 2 bits, then the best the ADC can provide is less than 32,000:1 SNR per scan. This represents the limit of state-of-the-art A/D performance, and to obtain an infrared spectrum with a greater SNR than ($32,000/\sqrt{1,000} \simeq 1000$:1) with current technology, one must go to signal averaging. In signal averaging, the SNR in the spectrum is improved by time averaging the random noise in each interferogram. The trading rules suggest that the SNR can be improved by a factor of 2 for a fourfold increase in measurement time. That is, the signal *increases* as a function of N, where N is the number of scans, and the noise in the spectrum increases as a function of $\sqrt{N}$, assuming the noise is truly random. Thus, the SNR increases as a function of $\sqrt{N}$. In practice, it is somewhat less than this, since not all sources of noise are random, and secondly, many commercial spectrometers have a SNR limit which they cannot exceed. In these systems, taking more scans can sometimes be detrimental to the quality of the spectrum. To return to the original example, at least 2000 scans are required to achieve the SNR required for many of the more-demanding problems.

To sum it up, aqueous-based biomedical FT-IR measurements are some of the most demanding studies in terms of required dynamic range. The high intrinsic absorbance of the aqueous medium, coupled with the very low intensities of most of the useful amide-related bands in proteins, for example, requires extremely high SNRs in the parent spectrum. Further mathematical operations, such as Fourier deconvolutions, require yet a higher SNR to allow successful completion of these manipulations.

III. DATA PROCESSING

A. Data Systems

One of the principal respects in which commercial FT-IR systems differ from one another is in the utility of the dedicated computer included as part of the FT-IR system. In general, one of two approaches has been taken: (1) specifically design a computer for FT-IR-type data processing; or (2) take an industry-standard mini- or microcomputer and add software and hardware support to make it appropriate for FT-IR data processing. In the past, the former approach was more common (with the exception of Digilab instrumentation). That is, specifically designed computers were used as the FT-IR data system. These computers have the advantages of being optimized for Fourier transforms, of having word lengths which were thought to be appropriate for the dynamic range of the interferogram, and of being efficient at signal averaging as the spectrometer is scanned. More recently, the trend has been to take industry-standard computer hardware and add interfaces and software to make them acceptable for the FT-IR application. Both approaches have merit, but it is the opinion of the author that the use of industry-standard computers is the preferred option, especially if the ultimate user of the FT-IR system wishes to do programming or to use the computer for functions other than signal averaging and data processing. The industry-standard computers have, of course, much more software and flexibility built into them. In the last year or so, there have been a number of FT-IR data systems announced which are based on standard microprocessors such as the Motorola 68000, and these data systems are general-purpose laboratory computers which can be used to collect FT-IR spectra. Vendors including Mattson Instruments, Digilab, Perkin-Elmer, and Beckman have general-purpose microcomputer systems that can be used to operate their respective FT-IR systems.

It is the author's opinion that array processors are highly desirable and should be considered if they are available for the computer that comes with the FT-IR. The presence of an array processor can speed up the Fourier transform procedure by as much as 100-fold, and it also frees the main computer CPU for other functions. Array processors are designed for parallel processing, and they do many of the computations in a parallel pipe arrangement with hardware assistance. In biological work, one often collects large amounts of data, especially if one is interested in real-time studies, and the ability of a dedicated array processor-based computer system to keep up in real time with the Fourier transform and absorbance calculations is a great asset. All of the systems at the National Center have array processors where available.

Another trend in data processing is towards powerful stand-alone computers for some or all of the post-Fourier-transform data processing. One has the option of using either one of the above-mentioned general-purpose minicomputers that the vendors are currently providing, or an industry-standard minicomputer such as the Dec/VAX line or the Data General Eclipse line. The advantages of having a powerful stand-alone computer are severalfold:

1. Multiple users can access a common spectral information base.
2. Specialized software programs and procedures can be more easily written and implemented on a general-purpose computer.
3. The total computer power of the stand-alone system is often considerably greater than that of the dedicated computers of the FT-IR system, which makes the response quicker when doing numerical analysis-type programs.
4. The larger minicomputers are generally multiuser and multitasking, and several operators can be reviewing spectra, plotting, and copying or storing spectra simultaneously. In general, this will increase overall efficiency. The ideal, of course, would be to have the dedicated computer of the FT-IR communicating to a stand-alone, more powerful minicomputer which could be used for subsequent data processing. For the typical user who does not have this luxury, one should consider purchase of a data system from one of the vendors who offer an industry-standard microcomputer. The general versatility and programmability of the industry-standard computer should be an asset.

B. Software Data Reduction of a Typical Biomedical ATR Experiment

To begin a typical biomedical experiment, we first align an Attenuated Total Reflection (ATR) flow cell which has been assembled and leak tested, allow the instrument to purge so that atmospheric water vapor is removed, and obtain a reference background of the cell without water. This is stored on the disk for later use. The cell is then filled with buffered saline (for aqueous experiments), and again a reference spectrum of the cell filled with water is collected and stored to disk. An absorbance spectrum of the water-filled ATR cell (the ATR crystal may or may not have been coated with a polymer film) is created by ratioing against the empty cell background. The instrument is put into the high-speed data-collection mode, which is often the same as the gas chromatography software mode, and rapid data collection is initated. Simultaneously, a valve is rotated which directs the blood or protein solution to the flow cell. Initially between 1 and 10 spectra per second are collected to disk, often with an array processor Fourier transforming and displaying the resultant spectra in real time. The computer at this stage is programmed for a direct ratio of the reference liquid-filled cell to the sample cell with protein, which is in effect a 1:1 water subtraction, and the result of this ratio is stored to disk in absorbance. Once the experiment has been terminated, we begin data processing on up to 500 infrared spectra. In the future we expect to transmit the raw data files to a stand-alone computer system for subsequent data processing. Currently, these processing steps are carried out on the dedicated FT-IR computer. The first data-processing task is to repeat and scale the spectral subtractions to adjust for the fact that as a protein layer builds up at the surface, less water appears to be present within the evanescent field of the ATR crystal, as evidenced by changing $1640/1550$ cm^{-1} band ratios. Subtractions are performed and several criteria are used to select the best subtraction factor. These criteria are discussed in a later section, but in summary, when dealing with proteins, we subtract to a fixed Amide I:II ratio which is known from D_2O and dry protein experiments, and secondly, we look for dispersion curves or negative bands on the high-frequency baseline above Amide I. Figure 2 shows an example of an over- , under- , and appropriately subtracted protein spectrum. Once the subtractions have been carried out, the computer measures baseline band intensity and frequency, and plots them as a function of time. We plot the baseline-corrected intensity of certain bands as indicators of the total amount of protein

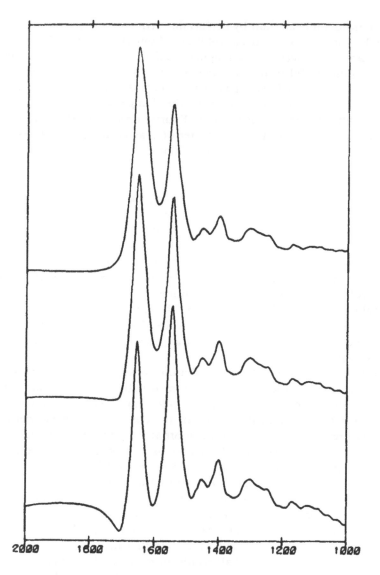

FIGURE 2. Two percent human serum albumin after water subtraction. Top, subtraction factor 0.98 (undersubtracted); middle, subtraction factor 1.00 (appropriately subtracted); bottom: subtraction factor 1.02 (oversubtracted). Note the dispersion curve at $\sim$1720 cm^{-1} in the oversubtracted spectrum.

adsorbed to the surface, and this information is used to track certain functional group-related bands which arise from components in blood such as glycoproteins. The next data-processing step involves bulk correction. We know that when adsorption experiments are carried out, the evanescent field penetrates into the bulk solution, and proteins in this bulk flowing layer are detected by the evanescent field. By using water as an internal standard, it is theoretically possible to calculate the contribution of bulk solution proteins to the total spectrum that is obtained experimentally. We are currently evaluating different correction factors which use the 1640 cm^{-1} water band as an internal standard. Once the best algorithm has been selected and verified, we anticipate that computers will routinely calculate the bulk contribution factor, and subtract transmission spectra of the same solution to compensate for that bulk contribution.[2] The next step in data processing depends on the particular nature of the experiment. One can learn a considerable amount by performing sequential subtractions of one file vs. another, and by using Fourier deconvolutions. Fourier deconvolution provides much information about changes in overlapping band ratios and intensities, and this has proven to be quite useful in tracking the changes in proteins adsorbed to a surface. Spectral subtraction of different spectra represents different times in the adsorption process, and this provides a direct probe of the changes which occur at the surface as a function of time. When carrying out the Fourier deconvolution step, it is often advisable to go back and rerun the peak center of gravity and peak intensity programs, getting new plots of band intensities and positions.

C. Fourier Deconvolution

In recent years a series of papers by Kauppinen, Cameron, Mantsch, and Moffatt[3,4] have illustrated that the process known as Fourier self-deconvolution is useful for data reduction in several classes of biomedical experiments. Fourier self-deconvolution causes computerized narrowing of infrared absorption bands, resulting in resolution of multiple bands which would otherwise be inherently overlapped. Fourier deconvolution is used when the intrinsic half-widths of the infrared bands are such that they overlap regardless of the instrumental resolution.

When dealing with a number of overlapped broad bands, which is generally the case with proteins and other complex macromolecules, Fourier deconvolution allows determination of the number of bands in an overlapped situation, and exact frequencies for each of those bands.

To appreciate how Fourier self-deconvolution works, recall the presentation in Table 1 which illustrates the relationships between information in Fourier space and spectral space. Specifically, information relative to band half-width in spectral space is expressed as the rate of exponential decay of the cosine wave in Fourier space. As a result, the broader the infrared band, the more rapidly the cosine wave decays. If one wished to mathematically narrow infrared bands of a specific width, one would need to reverse the exponential decay of the cosine waves associated with those bands. This is how the Fourier self-deconvolution procedure functions.

The Fourier deconvolution algorithm takes an infrared spectrum, generally after subtractions have been performed, and executes a reverse Fourier transform to return the spectrum to the Fourier domain. The operator is prompted for the degree of band narrowing desired, and the band half-width rate that the algorithm will be optimized for. When the user selects a band half-width for optimum band narrowing, one essentially informs the computer which rate of exponential decay is to be reversed. In other words, the interferogram will be multiplied by an increasing exponential function which is the inverse of that which would give rise to a band with the input half-width. The net result is cosine waves which no longer decay as fast as they did originally, and when this synthetic interferogram is Fourier transformed back to spectral space, the result is a spectrum in which the bands have been narrowed. One never gets something for nothing, and the deconvolution procedure is no exception. First, the program generally assumes a Lorentzian profile and one sacrifices information about band half-width and band shape to learn something about band position. Secondly, since the algorithm can only be optimized for bands of one specific half-width, there is considerable generation of artifacts when bands of differing half-widths are nearby in the same spectrum. This is perhaps the greatest problem with Fourier deconvolution — artifacts due to closely spaced bands with differing half-widths. And finally, multiplying an interferogram by an increasing exponential has a very drastic (negative) effect on the SNR. For every factor of 2 enhancement in the apparent resolution, that is, for a factor of 2 decrease in band width, roughly a factor of 10 in SNR must be sacrificed. It becomes apparent that very high SNR spectra are necessary before Fourier deconvolution can be used.

In defense of Fourier deconvolution, we have found it to be an extremely useful tool for studies of complex macromolecular systems. Most of the systems we are interested in have significant degrees of intrinsic band overlap, and Fourier deconvolution provides a way to accurately measure the number and locations of bands once one is assured that significant artifacts are not present. It is interesting to note that mathematically, second and fourth derivative algorithms are different weighting factors in Fourier space of a deconvolution-type procedure. Therefore, the information one gains relative to peak position by using derivative spectroscopy is quite similar to the information learned from Fourier deconvolution, and the two techniques should be viewed as complimentary.

The author views Fourier deconvolution as the first in a series of sophisticated algorithms which will be applied to the general problem of computer-assisted spectral interpretation and analysis, and an expanded use of the computer in spectral analysis will be essential in the future development of biomedical FT-IR.

D. Summary

As was mentioned in the previous section, it is the opinion of the author that the majority of the progress that will be made in interpretation and quantitation of biomedical infrared spectra in the next 5 to 10 years will be based on improvements and developments in computerized data processing. The FT-IR hardware itself can be expected to continue to improve (and will be discussed later in this chapter), but the more urgent need is for the computer to assist in interpretation. Biomedical FT-IR needs to develop to the point where one need not be a dedicated specialist to carry out the experiments and interpret the results. The development of algorithms for quantitative determinations, of pattern recognition techniques, for spectral interpretation and of computerized approaches for quantitation will all make a large contribution to the general utility of biomedical FT-IR over the next few years. As computer algorithms do more for us in terms of data processing, less bias will be put into the interpretation by the spectroscopist, and more information will be learned from the infrared spectra that are collected. As was mentioned at the beginning of this chapter, the infrared spectrum is information rich, and often there is more information than can be used because of the nonspecific nature of the infrared spectrum. However, it appears that with a number of appropriate data-processing techniques, much of the information that is currently present but unusable will be obtained. I personally look for a number of significant developments in the applications of computers to biomedical FT-IR over the next few years.

IV. FUTURE INSTRUMENTATION DEVELOPMENTS

FT-IR has undergone a great number of significant improvements in terms of both hardware and software performance over the last 15 years. The systems that were first available in the early 1970s had performance only marginally better than grating spectrometers, and the data systems were primitive. Advances in interferometer stability, scan speed, dynamic range, and SNR per scan as well as improvements in detector stability, D^*, and dynamic range, and improved sources and associated electronics have all contributed to improved instrumental performance of FT-IR.

On the data-processing side, the revolution in the electronics industry and the advent of powerful microprocessors and chip technology have caused a great revolution. Today's data systems are much more powerful, faster, smaller, and have opened whole new areas of applications for FT-IR.

As the importance of and the coming development of data processing have been stressed in the previous sections, I would like to focus this section on a brief discussion of what can be expected in terms of hardware development over the next few years.

When considering the future of IR, the first question is whether tunable lasers will have a major impact in the mid-infrared region. A tunable laser has a number of advantages such as coherent radiation, very high SNRs due to the high input power of the laser, and extremely high resolution due to the monochromatic nature of the light source. However, for biomedical applications, tunable lasers are probably not the answer. The high power of the laser would not be usable with biological systems that are subject to degradation with high input powers. There are likely to be problems in tuning the laser over a broad frequency range in general and in tuning the laser over a broad range very quickly making the high resolution most probably useless; finally, most problems today are not throughput related. Although tunable lasers will certainly be coming, they are likely to have a less significant impact in the biological arena as compared to other arenas.

Future advances that are likely to benefit biomedical FT-IR will primarily involve the A/D electronics and the detector itself. The most significant problem biomedical FT-IR faces today is the dynamic range problem. An 18- or 19-bit A/D converter with the same precision as that obtained from 15-bit A/Ds, coupled with a detector which can provide a signal with a dynamic range of this magnitude, would be a great improvement. As was mentioned earlier, the dynamic range requirements for interferograms are much more severe than those for spectra, and improvement in this area is very important to the future of FT-IR. There is reason to believe that better infrared detectors are in existence today in classified applications, and a number of electronics firms are working on improved A/D converters with the required dynamic range, precision, and speed.

The other major area of development will be in increased interferometer stability, both short- and long-term. As we develop hotter sources, lower intrinsic noise detectors, more rapid scan speeds, and greater dynamic range electronics, the band absorbances that will be routinely measured and quantitated will become less and less. As we measure less and less intense bands, baseline stability becomes an increasingly important factor. Several approaches are being taken to enhance baseline stability, including laser fringe counting as a reference for start of mirror scan, dynamic mechanical alignment such as the Bomem system uses, the use of corner cubes in the optics (Mattson Instruments) to decrease the problem with moving mirror tilt during scan, and more sophisticated scan algorithms which cause the mirror to either scan bidirectionally or to have different turnaround profiles on retrace.

I expect to see considerable improvement in the dynamic range of detectors and electronics over the next 3 to 5 years. Whether similar improvements in instrument stability will follow is an open question at this point. The vendors have noticed that there are more and more applications for low resolution, very high SNR spectrometers, and a number of them are working towards improving the baseline stability, scan speed, and SNR per unit time on instruments optimized for high SNRs. It is my belief that in the next 3 to 5 years we will witness a factor of 5 increase in SNR obtainable per unit time, primarily as a result of new detector and A/D electronics. I expect to see a factor of 2 to 3 improvement in baseline stability due to maneuvers such as corner cubes and dynamic alignment systems to maintain mirror stability during a scan.

V. HOW TO SELECT AN FT-IR FOR BIOMEDICAL USE

A. Instrumental Factors

As was discussed previously in this chapter, the primary consideration in the majority of biomedical experiments is the SNR achieved per unit time. General instrument characteristics which meet these requirements are high throughput, high SNR per scan, relatively fast scan rates, and well-optimized MCT electronics. On the other hand, since resolution is traded off at the expense of SNR, high-resolution instruments per se are not only unnecessary, but are often undesirable because compromises have been made in that direction. Biomedical experimentation requires a sound understanding of the factors which contribute to the SNR achieved in an experiment:

1. Resolution. SNR will be improved proportionally for every decrease in instrumental resolution; this author recommends 8-wavenumber resolution when possible, and at the most 4-wavenumber resolution in specific cases.
2. MCT detectors. Most aqueous experiments will benefit greatly from the use of MCT detectors. Careful selection of detectors optimized for a particular experiment can improve performance by two- to threefold.
3. Apodization. In keeping with the trend towards running at as low a resolution as possible, use of triangular or other low-resolution apodization functions, coupled with one or two levels of zero filling, is highly recommended.
4. Electronics. Great lengths should be undertaken to have the best available electronics in terms of MCT preamplifier, gain-ranging electronics (if used), and state-of-the-art A/D conversion with appropriate filters. Instruments can vary two- to threefold simply on the basis of the quality of their signal-processing electronics and their A/D conversion electronics. Careful evaluation of instruments will reveal these differences.
5. Measurement time. SNR increases with the square root of measurement time. To improve the SNR by twofold under otherwise constant conditions, the number of scans must be increased fourfold. The instrument should have a demonstrated capability to signal average properly.
6. Throughput. Within limits determined by the detector and A/D converter, increasing the total amount of energy reaching the detector will increase the SNR proportionally. Careful evaluation of instrument design and optical accessories selection is important here.
7. Optical filtering. If only part of the spectral range is of interest, use of optical band-pass filters to restrict the wavelength range seen by the detector can help somewhat by decreasing the dynamic range of the interferogram.

While SNR is by far the most important consideration, there is also a significant requirement for baseline stability. The high SNR provides the ability to measure very small optical absorbance differences, often in the range of 0.1 mAU or less. The ability to accurately measure band ratios and to quantitate peaks is often dependent on baseline changes. A drifting baseline, particularly one which has a sinusoidal-type appearance to it, can be disastrous in many of these very demanding experiments. Factors such as temperature stability of the instrument, air flow rate in the gas purge, room vibration, source temperature changes, and so on, all impact stability. Vendors can also select instruments off their own production lines which vary in stability considerably. Careful attention to stability factors such as those listed above, as well as maintenance of good alignment of the beam splitter and the interferometer, will all contribute to improved baseline performance.

Data reduction and computerized data handling are becoming increasingly important tools in biomedical spectroscopy. As the ability of the spectroscopist to probe a molecular system in more detail increases, the demands placed on the data system also increase in an almost exponential fashion. In this author's experience, inherent programmability of the dedicated data system, in addition to the ability of the data system to handle large numbers of Fourier transforms and mathematical operations efficiently, are extremely important. Recommendations as to the minimum data system for a serious biomedical FT-IR experimenter might be as follows:

1.   At least 10 Mbytes of on-line hard disk storage; up to 200 to 400 Mbytes could be used quite readily.
2.   The ability to perform Fourier transformations in real time (with Fourier transform times less than 1 sec) coupled with multitasking on the computer can greatly increase the productivity of the operator. Many vendors are currently moving towards this goal.
3.   Since these experiments generate tremendous amounts of data, a tape drive for archival storage is recommended.
4.   If multiuser and multitasking capabilities are available options with the instrument selected, these are highly recommended. This permits one to undertake data reduction on previous experiments while data are collected for the next. The multitasking nature of the data system can be a significant advantage and increase the overall throughput of the instrument.
5.   Data collects of the high-speed gas chromatography style are essential. The ability to collect several spectra per second is very useful when one is trying to study the rearrangements in the deposition of macromolecules onto surfaces, for example. A minimum speed of at least one spectrum every 2 sec is probably necessary, and if SNR permits, up to 10 to 20 spectra per second could be useful under certain conditions.

B. Evaluation and Selection

This section is intended to present the author's views on how one should go about selecting an FT-IR system to be primarily used for biomedical experimentation. These recommendations are based on the author's own experience in evaluating a number of competitive systems for applications involving primarily aqueous ATR experiments and aqueous transmission experiments.

First, if there is one primary type of experiment you will be doing, carry out the experiment on each system if possible. That is, do not let the vendor demonstrate his favorite experiment. Do your own. When we were evaluating systems for the National Center for Biomedical Infrared Spectroscopy, we handcarried a complete ATR accessory and liquid cell with us to each of the vendor sites. With some modifications, we were able to use this accessory with virtually all the competitive systems, and we carried out the exact same experiment on each of the vendor's systems. This gave us a tremendous amount of feedback about versatility of the system, alignment ease, and throughput of the instrument on a specific type of experiment, and it provided us feedback on how convenient or inconvenient it was to perform data reduction on that particular data system. While carrying a complete experimental set-up to various demonstration laboratories is inconvenient, it provides much more information as to which instrument is most suited to your particular needs.

Second, be present for the evaluations yourself. Get a feel for whether the technical staff of the vendor knows how to do your experiment properly and whether they seem comfortable or uncomfortable having their system carry out that particular type of experiment. This tells you two things: (1) if the vendor doing the demonstration is uncomfortable doing your experiment, the results may be compromised simply due to poor experimental approach; and (2) it will give you an idea of how difficult it is to carry out this experiment on the vendor's instrumentation.

Third, be aware of the parameters selected by the demonstration laboratory. Better yet, specify for them the resolution, apodization, gain ranging, zero filling, smoothing (there should be no smoothing), time of acquisition, any baseline correction, any scan correlation or rejection, and so on. Each of these parameters can affect the relative appearance of the data obtained, and it will otherwise be impossible to directly compare data collected on competitive instruments. In practice one should try to specify all of the collection parameters so that, again, the vendor operates under conditions that the potential purchaser is most comfortable with, as opposed to those that suit the vendor.

Finally, it is highly desirable to use the same detector for all evaluations. This is often not possible, and if it is not, then obtain information from the vendor about the particular D* and wavelength cutoffs of the detector used. One has to be careful that instrument comparisons are not simply comparisons of different MCT performances. While one would think that competitive vendors have detectors of similar performance, there is often considerable variation, and problems can be avoided by seeing that similar detectors are used. In our case, we carried our own MCT detector to each vendor and had them interface it to their electronics. This is the best controlled case for comparison of instrument performance.

Once the data were collected and it was time to evaluate the competitive systems, we used two principal criteria. First and foremost was the SNR per unit time obtained using an aqueous ATR setup. We took consecutive spectra which had been collected with water circulating through an ATR cell, and ratioed those against each other directly. This was converted to absorbance, and the zero absorbance noise signal after water bands and protein bands had been ratioed out was measured. This noise level was used as an index of the SNR per unit time that the particular instrument delivered. The next criterion was to take a 300- to 400-wavenumber spectral range and to measure the baseline tilt associated with that region. We took the 1600- to 1200-cm^{-1} region and measured the maximum tilt between the high and low data points in that region. Ideally, one would expect a straight line with noise superimposed, and any deviation or tilt was compared from instrument to instrument and found to vary considerably.

It is still a fairly common practice to negotiate with the hardware vendors to obtain custom-designed or selected FT-IR systems. When a specialized system is ordered, it becomes important to specify performance criteria in the purchase order. Even if the system ordered is a standard configuration, purchase-order specifications are still an excellent idea. If a system is being specified especially for biological-type work, I recommend specification of the following parameters:

1. The RMS SNR obtained under defined scan conditions. In our case, we specify the maximum noise on the absorbance scale that should be obtained in 1 sec of measurement time, through a germanium ATR crystal immersed in water, is between 1600 and 1200 wavenumbers.

2. The maximum baseline tilt allowable. We specify the maximum allowable baseline tilt in the 1600- to 1200-wavenumber region over a measurement time ranging from 1 sec to several minutes, using an ATR flow cell. By ratioing consecutive collected spectra with different times of collection, short- and long-term drift in the instrumental baseline can be detected.

3. Instrumental scan speed. We specify how many scans per second the instrument must be able to measure.

4. Allowable electronic noise level. When the infrared beam is totally blocked or the detector is disconnected, the intrinsic noise level of the electronics can be measured. It should be no more than 2 bit steps on a 16-bit A/D converter.

5. Software performance. Certainly specify that there be software capable of dealing with data collected at a 1 sec per spectrum rate or faster.

6.   Scan rejection rate. If the particular vendor's instrument uses correlation or other software to throw out bad scans, specify the percentage of scan rejection that is allowable. Rejecting too many scans is an indication of an instrument that is not very stable, and by specifying a low rejection rate, higher stability is ensured.

All in all, if enough parameters are specified (and the vendor will agree to meet them), unhappy surprises can be avoided. I suggest tying these specifications to acceptance of the instrument, and leaving a portion of the bill unpaid until the specifications have been achieved under mutually agreeable conditions.

ACKNOWLEDGMENTS

This work was supported in part by the National Institutes of Health under Grant No. HL-24015 and Grant No. RR-01367.

REFERENCES

1. Griffiths, P. R., *Chemical Infrared Fourier Transform Spectroscopy,* Interscience, New York, 1975.
2. Fink, D. J. and Gendreau, R. M., Quantitative surface studies of protein adsorption by infrared spectroscopy. I. Correction for bulk concentrations, *Anal. Biochem.*, 139, 140, 1984.
3. Kauppinen, J. K., Moffatt, D. J., Mantsch, H. H., and Cameron, D. G., *Appl. Spectrosc.*, 35(3), 271, 1981.
4. Cameron, D. G. and Moffatt, D. J., *J. Test. Eval.*, 12(2), 78, 1984.

Chapter 2

BIOMEDICAL FOURIER TRANSFORM INFRARED SPECTROSCOPY: APPLICATIONS TO PROTEINS

R. Michael Gendreau

TABLE OF CONTENTS

I. INTRODUCTION

In the previous chapter, the instrumental factors which influence the applicability of FT-IR to biomedical problems were reviewed. The next two chapters review experimental work that has been conducted with proteins, macromolecules, and lipids at interfaces and in membrane systems, respectively. These chapters attempt to illustrate general types and areas of application for this technology, and are not meant to be exhaustive reviews of the literature in these areas. Specifically, the application of infrared spectroscopy (IR) to the study of proteins and protein systems has a relatively long history compared to other biomedical IR applications, and the literature will not be discussed in depth in this work. Applications of IR to studies of proteins in aqueous solutions have a much more limited history, with work beginning in the mid 1960s. The aqueous work throughout the 1960s used conventional dispersive infrared spectrometers with limited frequency accuracy and signal-to-noise ratios, which made it very difficult to observe the weaker amide vibrations present in the proteins and polypeptides under study. Thus, it was the consensus of workers at that time that infrared protein spectra were for all intents and purposes very similar to one another, which is in fact the case, and that blood proteins, for example, would be indistinguishable in solution due to their great similarity, which has not proven to be the case. These studies gave way to FT-IR transmission studies of proteins in the early and mid 1970s. However, these initial FT-IR studies used relatively crude instrumentation by today's standards, and generally utilized TGS-type detectors. While early FT-IR systems had much greater frequency precision and reproducibility than instrumentation available prior to that time, FT-IR instruments of that day were still not delivering the signal-to-noise ratios which are needed to probe the subtle infrared spectral differences between proteins. Application of FT-IR systems utilizing MCT detectors to the study of various types of proteins and macromolecules in solution has been primarily carried out since the late 1970s by Jakobsen and Gendreau at Battelle, Cameron and co-workers at the NRCC, Cooper and Veis in the U.S., and Kellner in Austria.[1-5] Several conclusions have come out of these studies:

1.   Proteins do have unique infrared spectra. These spectra are complex, and the infrared bands are generally broad and highly overlapped. There is unique molecular-level information available which can be used to study factors related to conformational and composition changes in multicomponent systems.
2.   Interpretation of protein spectra is a very difficult task, and problems associated with water subtractions and with studying proteins under reproducible conditions, as well as the inherent difficulty (for the spectroscopist) of working with biological systems has made this a difficult area to work in. Nonetheless, excellent progress has been made in applying FT-IR to studies of proteins, and several dozen proteins have now been studied under reasonably physiological conditions.

II. SAMPLING AND OPTICS

A. Transmission

Transmission IR continues to be the most widely used and versatile sampling technique in FT-IR. As is probably familiar to the reader, the transmission experiment is extremely simple. A sample is placed in a holder mounted in the infrared beam so that infrared light is transmitted directly through the sample, and any absorbance of light by the sample is measured by a decrease in energy impinging on the detector due to the absorption of the sample. The transmission experiment is applicable to any sample that can be made thin enough, dilute enough, or that is inherently transparent enough so

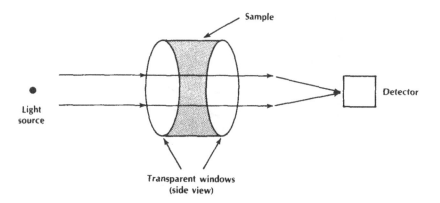

FIGURE 1.   Schematic of a typical mid-infrared liquid transmission cell.

that infrared energy may be transmitted through it. Typical transmission sampling techniques include the KBr pressed pellet, in which a solid sample is dispersed in a potassium bromide matrix and pressed into a pellet; the liquid cell, where a solution, including aqueous, is placed in a liquid-tight cell which has infrared transparent windows (such as calcium fluoride or potassium bromide), and light is allowed to pass through the liquid; and the gas cell, which is a transmission cell using gas-tight seals and infrared transparent windows. Gas cells tend to have much longer pathlengths than the cells used for liquids. A schematic of a typical mid-infrared liquid transmission cell is shown in Figure 1. Transmission is the simplest technique with which to undertake quantitative studies. Transmission experiments generally obey Beer's Law:

$$A = abc \tag{1}$$

where A = the infrared light absorbed (absorbance), a = the molar absorbance of the component, b = the pathlength, and c = the concentration of the absorbing component.

The infrared absorbance is linear with each of these factors, and Beer's Law is generally obeyed if there are no instrumental problems, if there is not a great deal of light scattering, and if the total absorbance of the analytical band does not exceed between 0.6 and 0.8 AU. At absorbances greater than roughly 0.8 AU, significant deviation from Beer's Law can be expected. For quantitative work, one should attempt to maintain band intensities at this level or below. The obvious trade-off is that if the absorbances of the strong bands are reduced by dilution or through use of shorter pathlengths, then all bands will be that much weaker, requiring lower total noise levels in the spectrum to measure the weaker bands. Recently it has been feasible to undertake quantitative transmission studies of biological molecules in water environments, in cells with pathlengths on the order of 8 to 10 μm using calcium fluoride windows. These studies have shown that the water absorbance band near 1640 wavenumbers can be mathematically subtracted and quantitative data on proteins obtained. Figure 2 shows a plot of baseline-corrected, Amide II peak intensities for two different proteins as a function of relative pathlength in a liquid cell. This figure is a plot of Amide II band intensity vs. measured water band intensity, and it illustrates that a quantitative relationship exists between water band intensity and protein band intensity at fixed concentrations.

B. Attenuated Total Reflectance

Attenuated total reflectance (ATR) is a highly versatile and useful sampling tech-

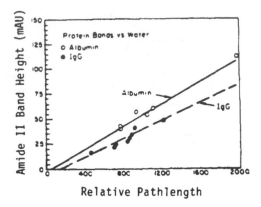

FIGURE 2.   Baseline-corrected Amide II peak band intensity vs. relative pathlength in a liquid cell for 1% human serum albumin and 1% IgG. This figure illustrates that a quantitative relationship exists between the water band intensity (x axis) and the Amide II band intensity (y axis).

nique for measuring samples on or near surfaces, and is an extremely useful alternative to transmission experiments. At first glance, ATR may seem somewhat mystifying. It takes advantage of the so-called evanescent wave, the standing electric field which exists at an interface when light is totally internally reflected. Total internal reflection occurs when light strikes an interface at which there is a change in refractive index, and when the angle at which the light strikes that interface exceeds the critical angle. When the critical angle is exceeded, total internal reflection takes place, and the light is reflected back into the first medium. An example of light exceeding the critical angle is the reflection of sunlight off a lake when the sun is low in the sky, the "mirror lake" effect. Quantum mechanical calculations show that when light is totally internally reflected at the interface, an "evanescent" wave is created at that interface, and this evanescent wave has an exponentially decreasing probability density as one moves from the interface into the second medium. The evanscent wave is required due to conservation of energy, and it is possible for this wave to interact with materials at that interface. If the evanescent wave can interact with materials at or near the interface, the next question is how far from the interface can materials be and still provide a contribution to the spectrum? The formula describing the so-called depth of penetration is from Harrick[6] and is given in Equation 2.

$$d_p = \frac{\lambda}{2\pi n_1 [\sin^2\theta - (n_2/n_1)^2]^{1/2}} \tag{2}$$

This expression for d_p relates that the depth of penetration, defined as the point at which the electric field of the evanescent wave has decreased to $1/e$ of its original intensity, is a function of the refractive index of the two media (n_1 and n_2), the wavelength of the light selected (λ), and the angle at which the ray of light struck the interface (θ). For germanium ATR crystals (refractive index $n_1 = 4.0$), an aqueous environment ($n_2 = 1.33$), and an incidence angle of 45°, penetration depths on the order of 0.1 λ are calculated, where λ is wavelength. For a typical protein band at approximately 6 μm, this penetration depth calculates to approximately 400 nm.

To consider quantitative work, we must derive a pseudo Beer's Law type of relationship for spectra obtained by ATR. This has been considered in two recent publications,[7,18] and a synopsis of those considerations is described following.

$$A^* = a^*bcfn \qquad (3)$$

where A^* = the ATR band absorbance, a^* = the ATR molar absorbance constant, b = the effective depth of penetration at the wavelength of the analytical band, c = molar solution concentration (film thickness with thin polymers), f = "effective surface area factor", and n = number of ATR reflections. These equations treat the depth of penetration similarly to the pathlength term in Beer's Law, concentration is a constant, and molar absorbance should also be a constant (although not necessarily the same as in transmission) if no polarization effects exist.*

There are several other factors in ATR which must be considered. The surface area of the crystal in contact with the second medium and the intimacy of the contact both affect the intensity of the measured infrared absorption band. The "effective surface area" term, f, corrects for these contact-related factors, and the number of reflections term, n, compensates for changes in ATR crystal geometry. By changing the geometry of the waveguide itself (i.e., by making it longer or thicker) the number of reflections an average ray of light makes going through the crystal can be increased or decreased. Since the evanescent wave is established at every reflection point, increasing the number of reflections has the exact same effect as increasing the surface area of the crystal itself. The difficulty in directly determining f has been one of the problems that has been experienced when ATR has been used for quantitative work, and generally an internal standard should be used so that the f and n terms may be eliminated from the computations. When dealing with aqueous systems, we generally regard the water band at 1640 cm^{-1} as an internal standard, and measurement of the water absorbance provides an empirical determination of the product of b, f, and n, since all are constant and are applicable to measurement of the water band, as well as the protein bands.

The other major consideration is the fact that the depth of penetration is a function of wavelength, and different bands will have different absorbances based on the wavelength at which they absorb and their distance from the surface. That is, for materials absorbing at longer wavelengths, they will have greater contributions at distances farther from the surface. If one considers a homogenous, infinitely thick material (i.e., greater than about 2000 nm in total thickness) which is not subject to polarization effects, then Equation 3 still holds.

So although a Beer's Law for ATR has more variables and requires considerably more assumptions, it is possible to conduct quantitative ATR on specific types of samples such as solutions. In an attempt to demonstrate this, a series of experiments were conducted using radiotagged (^{125}I) proteins. These proteins were allowed to adsorb onto a surface, and the absorbances of the Amide II and the 1400 cm^{-1} bands of the proteins were compared to the measured gamma activity of the surface, and it was found that the FT-IR data correlated with the measured radioactivity under controlled conditions.[8] While quantitative ATR is not as simple as quantitative infrared transmission, it is an achievable experimental result.

C. Flow Cell Design

Figure 3 shows the current flow cell used for aqueous ATR experimentation at the National Center for Biomedical Infrared Spectroscopy. This cell design is similar to previous designs that we and others have used, and it represents a compromise between optimal spectral performance and reasonable hydrodynamic flow conditions within the cell. This particular cell design is intended for studies of macromolecules; studies for which this cell was designed include enzyme interactions with substrates or inhibitors

---

* Recent experience in our laboratory has indicated that the molar adsorption constant (a) may not have the same value when going from transmission to ATR (a*). It is a constant value, but it may be difficult to predict the change when switching from transmission to ATR.

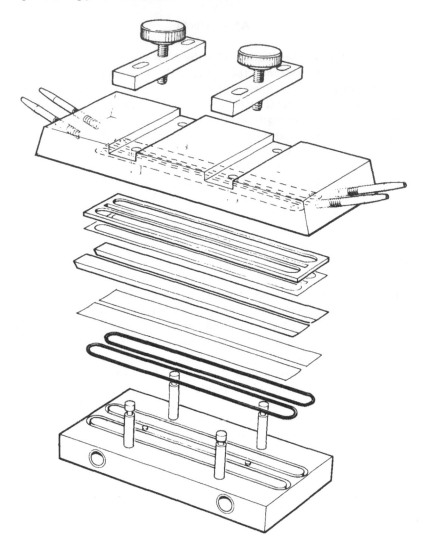

FIGURE 3.   Battelle-designed flow cell currently in use for most biological ATR experiments.

and protein deposition to surfaces. Biocompatibility studies in which whole blood is allowed to interact with polymeric surfaces coated onto the crystal have been one of the primary applications. This cell has several attractive features. It is a narrow-gap flow cell. The total height of the flow chamber is roughly 0.8 mm, allowing shear rates in the neighborhood of 500 to 700 sec^{-1} for volumetric flow rates in the neighborhood of 75 mℓ/min. The cell is relatively long with a narrowed aspect so that laminar-type flow rapidly develops inside the cell, and only the first 10 to 20% of the entrance region experiences turbulent flow conditions. The ATR crystal is 105 × 10 × 2 mm, and only one of the two available surfaces is utilized. Polymers of interest can be deposited on the active surface or the crystal can be precoated with other materials to study surface interactions. Flowing substrates or products are pumped into the cell to react or adsorb onto the coated or uncoated ATR crystal surface. The crystals themselves are germanium, 45° degree ATR crystals, designed to provide approximately 50 reflections while still allowing an energy throughout in the range of 10 to 15%. The cell is further designed to mount on a pair of guidebars which allow the cell to slide back and forth inside a focusing accessory. This provides for experiments in a dual channel configuration in which one channel can be used as a control for experiments performed in the other channel, or two separate experiments can be conducted simultaneously. This cell is the basis for the majority of the surface protein adsorption, biocompatibility, and enzyme-antigen denaturation and substrate studies that our group has carried out over the last 5 years. Related cells are in use at the Universities of Wisconsin and Utah, and a similar cell has been used for fluorescence studies at Stanford, Michigan, and Utah, and is discussed elsewhere in this volume.[9]

D. Aqueous ATR

A principal advantage of FT-IR over previous infrared techniques is that one can routinely work in an aqueous environment and study samples under more or less physiological conditions. The reasons that aqueous solutions do not pose a barrier to experiments today include the improved SNRs, the stability, and the reproducibility of the newer FT-IR spectrometers. All of these improvements contribute to an ability to digitally compensate for the absorption due to the water in aqueous solutions. By digitally subtracting away water, one mathematically produces a spectrum of the sample in the absence of water, although of course the sample is hydrated. The dilemma that immediately arises with digital water subtraction is knowing exactly how much to subtract. If one has the luxury of a sample with no absorption of its own in the neighborhood of 1640 cm⁻¹, then the subtraction is straightforward. Simply make the absorbance zero at that wavelength. However, most if not all interesting biomedical samples have absorbances in the region of the water band and as a result, subtractions are often complicated by underlying absorbances. Due to the inherent uncertainty of water subtraction in these cases, there are a number of approaches to performing the subtraction, none of which is fully satisfactory. An incomplete list of techniques currently in use is listed below:

1. No scaling. Simply assume that there has been no change in the intensity of the water band due to the addition of the sample, and perform a 1:1 subtraction using a previously collected water reference spectrum. This is satisfactory if there is no depletion of the water due to the presence of the sample, i.e., if there is no change in water concentration, and if the intermolecular hydrogen bonding of the water has not been disrupted by the sample.

2. Subtract to a fixed band ratio. This is a technique that we often use in our laboratories when working with proteins. Studies in the dry state and with D_2O have provided guidelines as to the expected 1640:1550 (Amide I:II) band ratio, and subtractions are scaled to provide this band ratio. This generally results in subtractions near 1:1, but often the amount of water subtracted is decreased by 1 to 2%, presumably due to the presence of a packed protein layer displacing water near the surface of the ATR crystal.

3. Subtract to achieve a flat baseline. Generally, and especially in the case of proteins, there is interaction between the water band and the Amide I vibration. If one oversubtracts, i.e., subtracts more water than is present and begins to subtract the Amide I band, a residual dispersion curve appears at roughly 1700 cm⁻¹. This dispersion curve appears as a dip that goes below baseline in absorbance, and one technique that has been partially successful is to attempt to maintain a straight baseline at frequencies above the water absorption band. This is a relatively straightforward approach, but it appears insensitive to small errors in water scaling.

4. Computerized subtractions. It is possible to automate the entire subtraction procedure using baseline criteria, band criteria, or center band frequency criteria, in which the sample band is at a slightly different frequency than the water band, and one measures the shift in band frequency as a function of water subtraction.

Each of these techniques can be automated by the computer. While each of these techniques is subject to its own errors, automating the subtraction procedure at least increases the reproducibility of the procedure. It is the author's opinion that perfect water subtractions in aqueous biological systems are not a realistic goal, since even if the exact subtraction factor were known, there is still the possibility of changes in hydrogen bonding and in water itself; one should therefore seek reproducibility.

Once a reproducible water subtraction has been obtained, one needs to correct for the depth of penetration function in the ATR experiment. Simple computer programs are now available which do this automatically. Since the depth of penetration is a linear function of wavelength, band intensities may be scaled in a linear fashion based on their wavelength and it is straightforward to correct the ATR spectrum. The result is generally a spectrum identical to that obtained through transmission.

III. PROTEIN STUDIES

A. Protein Transmission Studies

As a starting point for studies of proteins in solution, most investigators, including our own group, have initially used transmission as the method of choice. Aqueous solution protein spectra can be obtained in this way, and one has less concern about effects such as surface adsorption and protein denaturation which occur in ATR. In general, single protein studies were carried out, and this work has provided much of the basis for subsequent interpretation of ATR spectra. In transmission studies, physical paramaters such as concentration, pH, heat, ionic strength, etc., are easily varied, and this permits the induction of controllable changes in the protein. These induced differences can then be related to spectral changes. An example of this approach is given in Figure 4, which shows transmission spectra of varying concentrations of purified human serum albumin in saline solution at constant pH. Of significance are the changes with concentration occurring in the Amide III spectral region (1200 to 1350 cm^{-1}). (See Reference 10 for a more detailed discussion of the application of FT-IR to studies of protein spectra in aqueous solution.) The intensities of several pairs of absorption bands from Figure 4 were measured and ratioed, and the ratio was plotted as a function of concentration, shown in Figure 5. When the 1300/1250 cm^{-1} ratio is plotted, there is a gradual change in slope between 0.1 and 5% (weight/volume) concentration. Since infrared bands of the Amide III region are involved in the ratios plotted, and since previous infrared and Raman studies have indicated that the Amide III region is sensitive to conformational changes, the change in slope noted implies that a change in the organization or structure of the molecule occurs with changes in this range of solution albumin concentration.

The effects of varying pH are shown in Figure 6, which shows spectra obtained for solutions of albumin (5% w/v) in which the pH was varied from 3 to 9. In this figure, it can be seen that there is a change in the ratio of the 1450 cm^{-1} infrared band to the 1400 cm^{-1} band as the pH is lowered. By a pH of 3, the 1400 cm^{-1} band has almost disappeared. Koenig has assigned a Raman band at this frequency to the symmetric COO− stretching vibration of the carboxylate groups of amino acid side chains. This assignment is supported by the infrared data shown in Figure 6. For basic solutions (high pH) the carboxylate frequency (1400 cm^{-1}) is strong, but as the pH is lowered the carboxylate band disappears because the protonated form is generated. Also, with changing pH there are distinct changes in the Amide II intensity and in the Amide III spectral region involving an intensity increase in the 1250 cm^{-1} band as compared to the 1300 cm^{-1} band. This has been interpreted as an albumin conformational/subunit change at low pH, representing loss of α-helix.

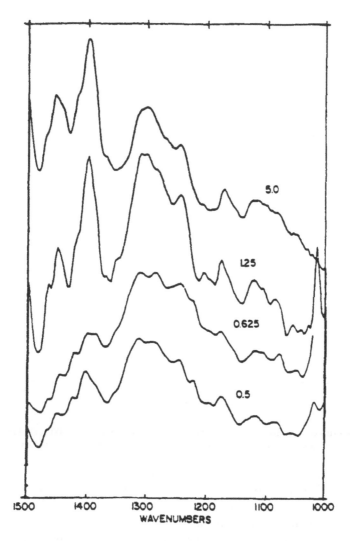

FIGURE 4. Transmission spectra of human serum albumin in saline solution. Concentration was varied from 0.5% w/v to 5.0% w/v (bottom to top). Spectra are water subtracted and plotted at different scale expansions to illustrate differences. (From Gendreau, R. M., Leininger, R. I., Winters, S., and Jakobsen, R. J., in *Biomaterials Interfacial Phenomena and Applications,* Cooper, S. L. and Peppas, N. A., Eds., ACS Adv. Chem. Ser. 199, American Chemical Society, Washington, D. C., 1982, 371. With permission.)

Transmission studies have been made of a number of other available proteins as solutions, films, and in KBr pellet matrices. These studies demonstrated that attempting to use Amide I frequencies for conformation assignment often leads to confusing results, but use of bands in both the Amide I and Amide III ranges is more often successful, especially if deconvoluted spectra are used to resolve the broad Amide I and Amide III infrared bands. Table 1 lists the solution Amide I frequencies for the observed conformations of the blood proteins, while Table 2 lists the solution Amide III frequencies for these compounds. All frequencies in Tables 1 and 2 come from deconvoluted spectra. In a number of cases deconvolution has proven highly useful for tracking subtle but reproducible spectral changes in protein molecules.

Note that in Table 1, some frequencies are assigned to two conformations (fibrino-

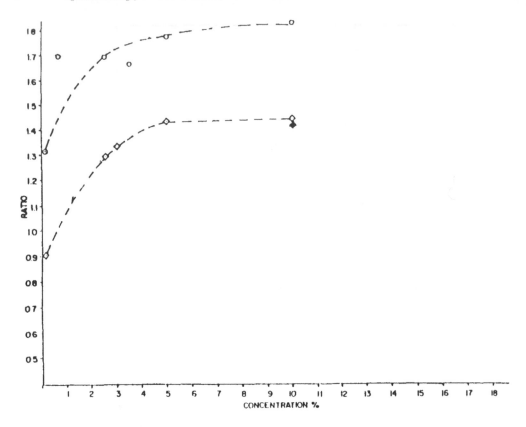

FIGURE 5.   The ratio of two infrared band pairs: 1300/1250 (◇) and 1400/1300 (○) plotted as a function of concentration in solution for human serum albumin. A change in ratio is indicated by a change in slope.

gen, hemoglobin) or that the frequency range for one conformation overlaps that of another (α-helix, 1670 to 1650 cm^{-1} and random or disordered, 1680 to 1640 cm^{-1}). By using the Amide III frequencies as well, these ambiguities often can be resolved, and the various conformations existing in each protein better defined.

For the Amide III frequencies, α-helix peaks occur over a range of frequencies near 1300 cm^{-1}, and some proteins exhibit several frequencies. Similarly, several frequencies have been observed for what is believed to be disordered structures. Definitive assignment of a number of frequencies in Table 2 to disordered structures is not possible at this time, and the possibility exists that some of these vibrations may be due to side-chain functional groups in the peptides themselves, or that some of these bands may arise from β-sheet structures. The Amide III region (1320 to 1230 cm^{-1}) certainly contains side-chain-derived bands as well as true Amide III bands, and differentiation of these is currently difficult, if not impossible. Structures such as β-turns[12] have also been identified which further complicate band assignments. The discussion in the following section describes some of the work that has been undertaken to assign bands in the Amide I region of a specific protein of known structure, ribonuclease.

B. Ribonuclease Studies

Ribonuclease has been extensively studied by a variety of physical methods and its crystal structure has been determined by X-ray diffraction.[13] This enzyme is therefore a good candidate for development of protein structure elucidation techniques. The deconvoluted Amide I region of ribonuclease has been interpreted in terms of protein secondary structure, and the assignments made are being supported by deuterium-exchange experiments which evaluate the degree of denaturation.

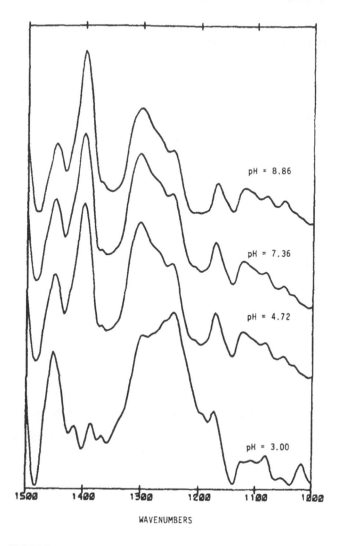

FIGURE 6.   Transmission spectra of human serum albumin in saline solution; pH was varied from 3.00 to 8.86 (bottom to top). Note the disappearance of the band near 1400 cm⁻¹ as the carboxylate side chains are protonated.

The Amide I regions of the Fourier deconvoluted and water-subtracted spectrum of ribonuclease in water are shown in Figure 7. The assignment of bands to types of secondary protein structure is made according to the literature:[14,15] (1) 1686 cm⁻¹, turns and β-sheet; (2) 1666 cm⁻¹, turns and regions of undefined structure; (3) 1655 cm⁻¹, α-helix; and (4) 1641 cm⁻¹, β-sheet. This interpretation is consistent with the content of β-sheet (40%) and α-helix (20%) found in the X-ray structure. The deconvoluted spectrum of ribonuclease in D_2O, shown in Figure 8, is similar to the H_2O spectrum except for the shifts in peak frequencies which are expected as a result of H-D exchange.

Tritium-exchange studies have indicated that certain amide protons of ribonuclease will not exchange with the solvent unless the enzyme is heated and unfolded.[16] These protons are thought to be in "structured" protein domains such as α-helical and β-sheet regions. Figure 9 shows the spectrum of a solution of ribonuclease in D_2O that has been heated for 30 min at 62°C, a condition at which complete amide proton exchange occurs. The positions of peaks thought to contain a large component of α-

Table 1

OBSERVED AMIDE I FREQUENCIES (H₂O SOLUTION) FOR VARIOUS BLOOD PROTEINS

Protein	Solution conc (wt. %)	Conformation frequencies (cm⁻¹)ᵃ		
		α-Helix	Random coil	β-Sheet
Albumin	1	1655ᵇ	1655 s, 1678	—
γ-Globulin	1	—	1670 m	1640 s
Fibrinogen	2	1650 s	1650 s, 1675 m	1630 Sh
β-Lactoglobulin	2	—	1605 m, 1680 Sh	1630 s
α₁-Acid glycoprotein	1	1665 Sh	1650 s	1630 s
Hemoglobin	1	1655 s	1655 s	1635 s
Thrombin	1	—	1645 s	—
Prothrombin	?	1660 m	1640 w	1620 s

ᵃ The frequencies are those of the deconvoluted spectra.
ᵇ s = Strongest band in Amide III region, m = medium band in Amide III region, w = weak band in Amide III region, and Sh = shoulder on another band.

Table 2

OBSERVED AMIDE III FREQUENCIES (H₂O SOLUTION) FOR VARIOUS BLOOD PROTEINS

Protein	Solutions conc (wt. %)	Conformation frequencies (cm⁻¹)ᵃ		
		α-Helix	Random coil	β-Sheet
Albumin	1	1350 s, brᵇ	1265 m, 1242 w	—
γ-Globulin	1	1312 m, 1284 m	—	1230 s
Fibrinogen	2	1280 m	1260 s 1249 s	1233 w
β-Lactoglobulin	2	1315 s, 1285 m	1261 w	1238 s
α₁-Acid glycoprotein	1	1320 s, 1298 w, 1280 m	1260 w	1241 m
Hemoglobin	1	1350 s	1265 s, 1245 s	1245 s
Thrombin	1	1295 m, 1280 s	1252 m	—
Prothrombin	?	1320 m, 1302 m, 1285 m	1261 s	1242 m

ᵃ The frequencies are those of the deconvoluted spectra.
ᵇ s = Strongest band in Amide III region, m = medium band in Amide III region, w = weak band in Amide III region, and br = broad band.

helix or β-sheet (a, c, and d) shift to lower wavenumbers after heating, a result consistent with further exchange of amide protons (Table 3). The peak assigned to turns and random structure (b), on the other hand, does not change its position. This was as expected, since these protons should have previously exchanged with deuterium while at room temperature. The combination of deconvolution and deuterium exchange studies has proven to be a useful way to confirm such spectral-structural correlations, and is providing the basis for band assignments in more complex systems.

C. Protein Mixture Analysis

The transmission studies of aqueous solutions of albumin and ribonuclease described previously demonstrate that spectral changes may be induced in single protein systems. It is obviously important to demonstrate that individual proteins can be identified in a mixture. This implies that the spectral differences between molecules must be greater than the range of spectral changes which occur in either isolated molecule. Techniques that have been used to identify proteins in mixtures include:

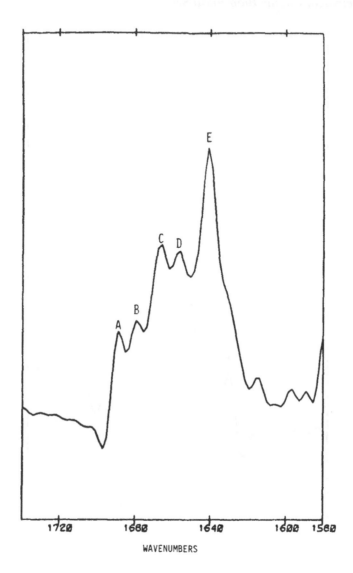

FIGURE 7.   Spectrum of Amide I region of 5% ribonuclease in H₂O.
This spectrum has had water subtracted and has been Fourier deconvolved.

1.   Spectral subtraction, in which reference spectra are mathematically subtracted from the mixture spectrum.
2.   Second- and fourth-derivative spectra, in which calculation of the derivative spectra helps to locate peak position, and thus resolve overlapping bands.
3.   Fourier deconvolution, in which computerized resolution enhancement is used to help locate band position and resolve overlapped bands.

Overlapping bands have been identified as a significant problem for biomedical FT-IR. Such overlap is generally intrinsic because the half-width of most bands in macromolecules is large. This implies that increasing the instrumental resolution of the spectrometer will not help resolve the bands, so other approaches need to be taken.

The use of spectral subtraction for mixture analysis is shown in Figure 10 in which a mixture of albumin and gamma globulins (1:1) is compared to spectra of the individual proteins. It would be difficult to directly identify either of the individual components

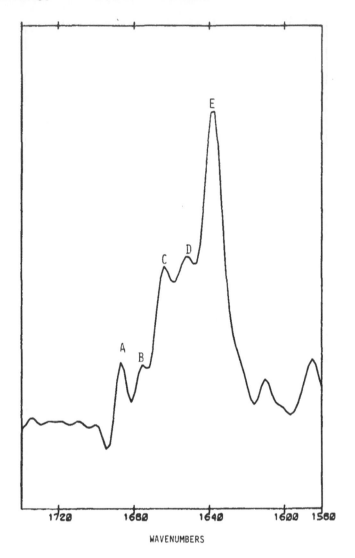

FIGURE 8.   Spectrum of Amide I region of 5% ribonuclease in D_2O after being held at 62°C for 0 min.

from the spectrum of the mixture. Figure 11 shows the result (bottom) of subtracting a spectrum of albumin from the spectrum of the mixture. This subtracted spectrum matches well with the reference spectrum of gamma globulins (top). The reverse subtraction is shown in Figure 12, which compares favorably with albumin.

The use of second-derivative spectra to identify proteins in mixtures is shown in Figure 13. At the top of the figure, a second-derivative spectrum of the 1:1 mixture of albumin and gamma globulins is compared to the absorbance spectrum of this mixture (bottom). Although there is a slight low-frequency asymmetry on the Amide I (1650 cm⁻¹) band, it would be difficult to identify the components. The second-derivative spectrum of the mixture shows four peaks (marked with arrows) in the Amide I region. The two high-frequency peaks correspond to the frequencies of two peaks observed in the second-derivative spectrum of pure albumin, while the two lower-frequency peaks of the mixture correspond to peaks seen in the second-derivative spectrum of gamma globulins. Typical literature values for albumin are 55% α-helix and 15% β-pleated sheet conformations.[17] The high-frequency Amide I peaks in the second-derivative

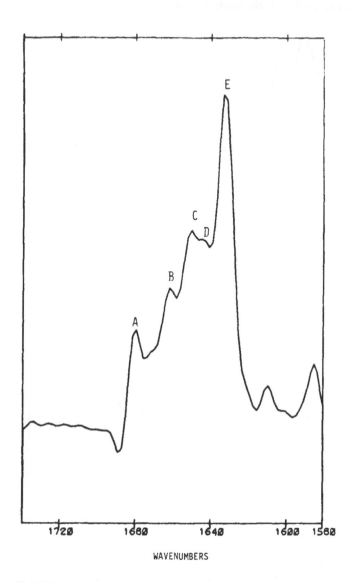

WAVENUMBERS

FIGURE 9. Spectrum of Amide I region of 5% ribonuclease in D₂O after being held at 62°C for 30 min. The 62°C allows proton exchange to occur.

Table 3
AMIDE PROTON EXCHANGE WITH DEUTERIUM

Time at 62°C (min)	Peak positions (cm⁻¹)			
	A	B	C	D
0	1687.7	1664.6	1653.0	1637.6
5	1683.9	1664.6	1651.1	1635.6
30	1680.0	1664.6	1649.1	1633.7

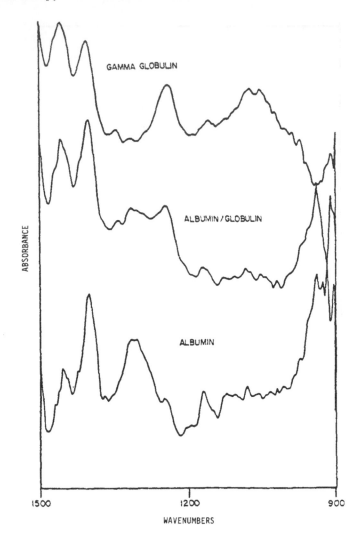

FIGURE 10.   Top, transmission spectrum of 1% w/v gamma glob-
ulin from 1500 to 900 cm⁻¹; bottom, human serum albumin; middle,
1:1 mixture of albumin and gamma globulin. (From Gendreau, R. M.,
Leininger, R. I., Winters, S., and Jakobsen, R. J., in *Biomaterials:
Interfacial Phenomena and Applications*, Cooper, S. L. and Peppas,
N. A., Eds., Adv. Chem. Ser. 199, American Chemical Society,
Washington, D. C., 1982, 371. With permission.)

spectrum of the mixture agree in frequency with the frequencies assigned to α-helix
(1655 cm⁻¹) and β-pleated sheet (1685 cm⁻¹) conformations of albumin.

Figure 14 illustrates application of Fourier deconvolution to this same mixture. This
figure shows the Amide I and Amide II region for an absorbance spectrum of a 1:1
albumin-gamma globulin mixture (top) and the deconvolution of this spectrum (bottom). In the deconvoluted spectrum (bottom), two peaks are clearly discernable at 1655
(albumin α-helix) and 1643 cm⁻¹ (gamma globulins β-sheet). These frequencies correspond to Amide I peak frequencies of albumin and gamma globulins, as was the case
with the derivative spectrum. The serious worker in biomedical FT-IR will find that
spectral subtraction, deconvolution, and second derivative techniques all provide information helpful in identifying proteins in mixtures. In general, other software approaches including center of gravity peak finding, band half-width measurements, use

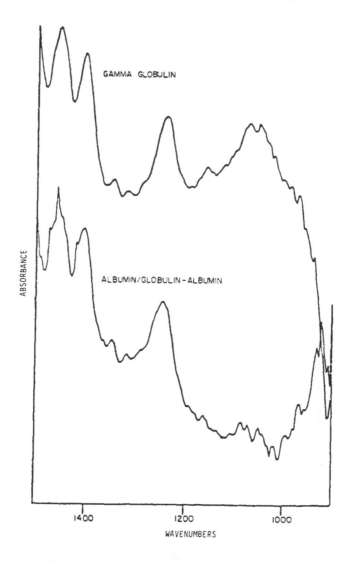

FIGURE 11. Result of subtraction of human serum albumin from
1:1 albumin-gamma globulin mixture (Figure 10, middle). Top, reference spectrum of gamma globulin (Figure 10, top); bottom, subtraction result (this subtraction should match the gamma globulin reference shown at top). (From Gendreau, R. M., Leininger, R. I.,
Winters, S., and Jakobsen, R. J., in *Biomaterials: Interfacial Phenomena and Applications,* Cooper, S. L. and Peppas, N. A., Eds.,
ACS Adv. Chem. Ser. 199, American Chemical Society, Washington,
D. C., 1982, 371. With permission.)

of multiple bands, and baseline correction are also necessary when working with complex systems.

The transmission FT-IR studies of aqueous protein solutions indicate how structural
and conformational differences in a protein can be related to spectral changes, and
that spectral features can be used to identify proteins in mixtures. These studies involve
static systems, and many of the potential biomedical applications involve studying dynamic processes, usually as they occur on a surface. Studies such as enzyme reactions,
antigen-antibody interactions, protein surface adsorption, cell-surface interactions and
the like require a technique compatible with flow and surface studies.

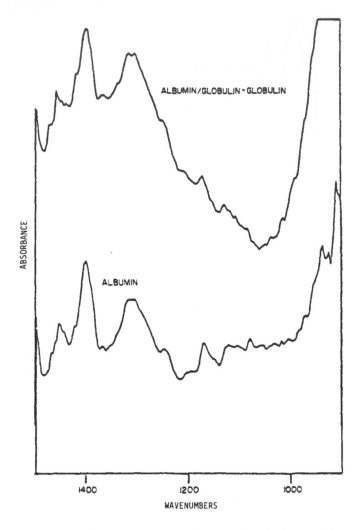

FIGURE 12.   Result of subtraction of gamma globulin from 1:1 albumin-gamma globulin mixture (Figure 10, middle). Top, subtraction
result (which should match the human serum albumin reference
shown below); bottom: reference spectrum of human serum albumin.
(From Gendreau, R. M. Leininger, R. I., Winters, S., and Jakobsen,
R. J., in *Biomaterials: Interfacial Phenomena and Applications*,
Cooper, S. L. and Peppas, N. A., Eds., ACS Adv. Chem. Ser. 199,
American Chemical Society, Washington, D. C., 1982, 371. With permission.)

D. Protein Adsorption Studies

For near surface studies, ATR is an excellent experimental tool, and represents one
of the primary advantages of FT-IR. When a flowing ATR experiment is carried out,
the surface of the ATR element may be continuously monitored (scanned) as protein
adsorbs. Quantitative changes in the amount of adsorbed proteins can be detected as
well as changes in the conformation of adsorbed species (when working with single
proteins), and this allows studies of the changes in the composition or organization of
an adsorbed layer.

This type of information can be obtained from more complex, flowing mixtures as
well. This is illustrated by consideration of the interaction of a 1:1 albumin-fibrinogen
mixture with the surface. Representative spectra of the protein layer obtained by flow-

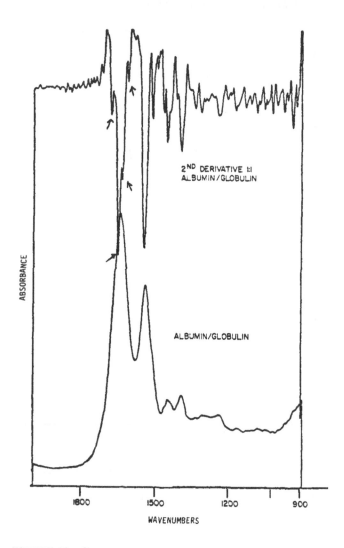

FIGURE 13.   Top, *second-derivative spectrum* of mixture shown in Figure 10; bottom, original mixture spectrum. Arrows mark four peaks also found in original mixture spectra. (From Gendreau, R. M., Leininger, R. I. Winters, S., and Jakobsen, R. J., in *Biomaterials: Interfacial Phenomena and Applications,* Cooper, S. L. and Peppas, N. A., Eds., ACS Adv. Chem. Ser. 199, American Chemical Society, Washington, D. C., 1982, 371. With permission.)

ing this mixture past a germanium ATR crystal are shown in Figure 15. It is important to note that in the Amide III spectral region, several changes occur with time of flow. In the series of infrared bands near 1300 cm^{-1}, there are changes in the number of bands, the frequencies of the bands, and the shape of the bands with time. The same type of behavior is observed for the infrared bands near 1250 cm^{-1}. A change in the intensity ratio of the 1300 cm^{-1} and 1250 cm^{-1} bands with time of flow is noted. All of these changes have been related to changes in the quantitative amounts of the two proteins on the surface. In general, proteins have been found to differ in their Amide III spectral features, and this region is useful for following changes in mixtures.

Because of the speed with which the FT-IR can collect spectra, intensities of various infrared bands can be measured nearly continuously, and these intensities plotted against time of flow. If infrared bands which are not very sensitive to conformation or

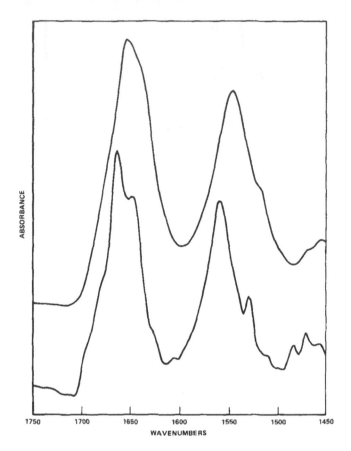

FIGURE 14.   Bottom, Fourier deconvolution spectrum of the mixture shown in Figure 10 (note appearance of bands which are now resolved). Top, original mixture spectrum. (From Gendreau, R. M., Leininger, R. I., Winters, S., and Jakobsen, R. J., in *Biomaterials: Interfacial Phenomena and Applications,* Cooper, S. L. and Peppas, N. A., Eds., ACS Adv. Chem. Ser. 199, American Chemical Society, Washington, D. C., 1982, 371. With permission.)

structural changes are selected (such as Amide II), measured band intensity can be related to quantitative amount of adsorbed material. Figure 16 illustrates this for the prior fibrinogen-albumin experiment. As can be seen, most of the protein diffuses into the boundary layer and adsorption occurs almost immediately. At slightly less than 2 min of flow, there is a decrease in the rate of protein arrival at the surface. Although after 2 min the rate of protein adsorption decreases, it never reached zero during the 222 min that the experiment was conducted, although only the first 42 min of data are shown in Figure 16.

If saline is substituted for the flowing protein solution, then kinetics of desorption can be studied as shown in Figure 17. Such a plot can be used to determine the ratio of stable adsorbed proteins to loosely bound and bulk solution proteins. Figure 17 shows that roughly 20% of the infrared band intensity is lost during the first 2 min of saline flow. This loss is due to bulk or loosely bound protein being removed, and after that time virtually no more protein is removed by flowing saline. Figure 17 indicates that the stable equilibrium layer upon saline wash has an intensity of about 27 mAU for the Amide II band. The intensity of the Amide II band reached 27 mAU at approximately 100 min of flow time. Thus, the adsorbed layer established by 100 min of flow in this

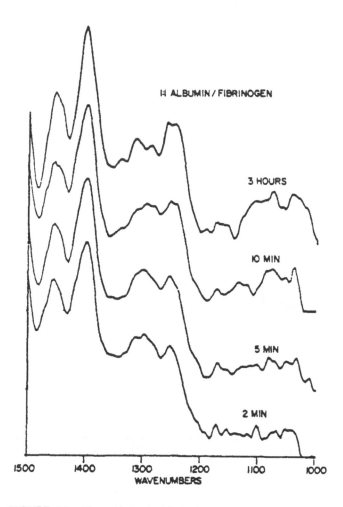

FIGURE 15. Illustration of changes in surface-adsorbed layer(s) which occur over time. The circulating solution was a 1:1 mixture of human serum albumin and human fibrinogen. Flow rate was 15 mℓ/min. This figure illustrates the layer(s) present at 2 min, 5 min, 10 min, and 3 hr (bottom to top, respectively). (From Gendreau, R. M., Leininger, R. I., Winters, S., and Jakobsen, R. J., in *Biomaterials: Interfacial Phenomena and Applications*, Cooper, S. L. and Peppas, N. A., Eds., ACS Adv. Chem. Ser. 199, American Chemical Society, Washington, D. C., 1982, 371. With permission.)

experiment is an amount of protein equal to the amount seen in the stable layer in the desorption experiment. Based on experiments using radiolabeled ^{125}I IgG, this infrared band intensity equates to approximately 0.25 $\mu g/cm^2$ of protein tightly adsorbed to the surface. To summarize our interpretation of the events which take place in a competitive experiment of this type as a function of time:

0—1 min	Mixture rich in albumin diffuses to the surface and adsorbs. This time interval is characterized by cell filling and diffusion of proteins to the surface. Albumin with a lower molecular weight diffuses more rapidly than fibrinogen.
1—7 min	Amount of surface-adsorbed albumin increases.
7—20 min	Relative amount of albumin at the surface begins to decrease. Because the total amount of protein is slowly increasing, there

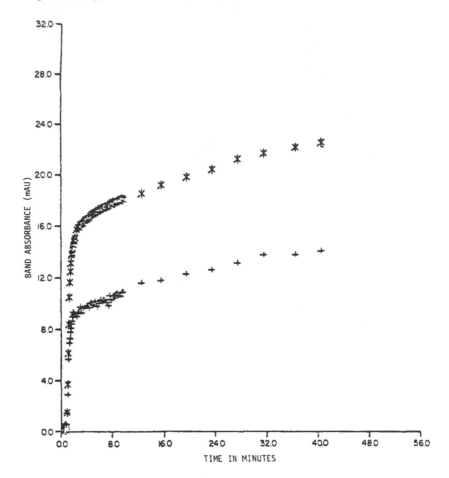

FIGURE 16.   The intensities of the Amide II band (✱) and the 1400 cm⁻¹ band (+) plotted as a function of time for the experiment described in Figure 15. These data were collected during the absorption phase. (From Gendreau, R. M., Leininger, R. I., Winters, S., and Jakobsen, R. J., in *Biomaterials: Interfacial Phenomena and Applications,* Cooper, S. L. and Peppas, N. A., Eds., ACS Adv. Chem. Ser. 199, American Chemical Society, Washington, D. C., 1982, 371. With permission.)

	must be replacement of albumin by fibrinogen. At 19 min, a fibrinogen conformation change may occur.
20—100 min	Albumin continues to be replaced. At 100 min, the amount of total absorbed material is equal to that seen in the stable desorbed layer (∼0.25 μg/cm²).
100—200 min	Albumin virtually disappears. By 200 min, the adsorbed layer appears to be nearly pure fibrinogen.

It would seem that FT-IR is a particularly useful method for studies of surface competition. Table 4 contains a summary of the information content present in protein spectra. Note that for Amide I and Amide II, the only strong bands, the Amide I frequency in general can only be used for qualitative purposes because of water-subtraction problems, and the Amide II band is generally useful for quantitative information. In most cases, the majority of information on protein identities and structure is obtained from the use of a series of very weak infrared bands between 1500 and 1000 cm⁻¹ in combination with Amide I and II. The need to study weak bands results in a

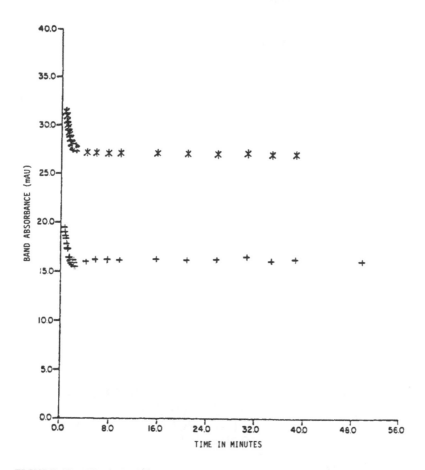

FIGURE 17. The intensities of the Amide II band (✖) and the 1400 cm⁻¹ band (+) plotted as a function of time for the experiment described in Figure 15. These data were collected during the desorption phase with saline circulating through the cell. (From Gendreau, R. M., Leininger, R. I., Winters, S., and Jakobsen, R. J., in *Biomaterials: Interfacial Phenomana and Applications*, Cooper, S. L. and Peppas, N. A., Eds., ACS Adv. Chem. Ser. 199, American Chemical Society, Washington, D. C., 1982, 371. With permission.)

corresponding instrumental sensitivity requirement, and this is the primary reason why FT-IR is being more successfully applied to biomedical studies today than in the past.

E. Protein Adsorption to Polymers

Using anticoagulant whole human blood, flowing ATR experiments using crystals which have been coated with films of various polymers have been compared.[16] Polymer films which have been studied include Biomer®, silicon rubber (SR), and poly(vinyl chloride) or PVC. These studies showed that in the adsorption process:

1. At 15 sec there are significant differences in the protein adsorption pattern among the three surfaces (Figure 18 — glycoprotein region near 1080 cm⁻¹ and absorption band near 1361 cm⁻¹).
2. After 1 min of flow, the differences between the three surfaces are less pronounced.
3. After 3 hr, differences still exist between the three surfaces, but these are minor compared to the differences observed at 15 sec (Figure 19).

Table 4

INFORMATION CONTENT IN PROTEIN SPECTRA

Band frequency (cm^{-1})	Source	Information content
1730 (1740—1710)	Lipid carbonyl	Associated lipids with proteins
1650 (1670—1630)	Amide I	Protein identification, presence of secondary structures, conformation changes
1550 (1570—1520)	Amide II	Protein quantitation, presence of multiple components, used for guiding water subtraction, Amide I/II ratio
1450	CH$_2$ sidechains? from valine, leucine, and isoleucine	Protein identifications
1400	Carboxylate sidechains	Protein quantitation pH sensitive
1361	Low-molecular-weight blood component	?
1300(1310—1240)	Amide III complex	Protein identifications, conformation changes, primary region used for differentiating proteins in mixtures; information relative to secondary structure is present as well
1200(1200—1160)	?	Characteristic bands for some proteins
1080 (1080—1040)	Carbohydrate functional groups	Glycoprotein content in adsorbed layers

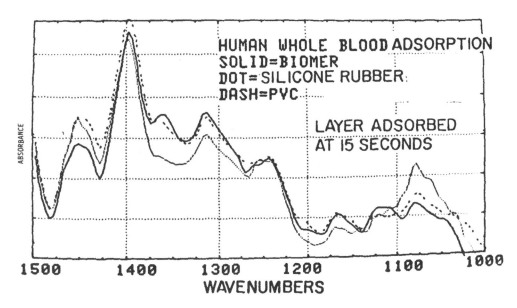

FIGURE 18.   Protein layer(s) adsorbed from human whole blood onto polymer surfaces after 15 sec of exposure. Three surfaces are shown: Biomer® (—); silicone rubber (····); and PVC (—·).

The differences at the various adsorption times may be related to complete surface coverage; i.e., until a monolayer is adsorbed, the protein sees the polymer surface and adsorption is affected by the differences between the polymers. Thus, the 15-sec spectra reflect protein diffusion, cell filling, and protein adsorption onto the polymer surface. After 1 min of adsorption time, the proteins in solution are likely seeing a surface primarily composed of other proteins. The nearly instantaneous appearance of a band located at 1361 cm^{-1} (Figure 20) was very interesting, and has led us to consider the possibility of a low-molecular-weight component which rapidly diffuses to the surface from whole blood. Attempts to identify this compound have established that this band

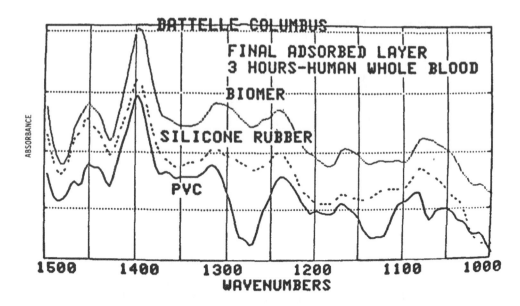

FIGURE 19. Protein layer(s) adsorbed from human whole blood onto polymer surfaces are shown: Biomer® (–··); silicone rubber (···); and PVC (——).

is due to a plasma component with a molecular weight of 2000 or less, and an inorganic bicarbonate is the most likely candidate. Whether this material actually surface adsorbs or is just concentrated in the boundary layer is not clear; however, a considerably larger amount of this species is seen in the ATR spectra than would be predicted based on transmission studies of the same blood or plasma sample, and is therefore of considerable interest.

An abbreviated summary of representative differences which have been observed spectrally between test polymer surfaces is listed below:

1. The composition of the initially adsorbed protein layer differs from one polymer to another. Biomer® displays most of the component giving rise to the 1361 cm⁻¹ species, while SR adsorbs more glycoprotein (1080, 1040 cm⁻¹) than Biomer® or PVC. All three polymers adsorb a mixture rich in albumin as compared to fibrinogen, but Biomer® adsorbs the most albumin and PVC the least.

2. The composition of the protein layer desorbing in the first few seconds differs from one polymer to another. Biomer® desorbs the 1361 cm⁻¹ species, Biomer® and SR desorb what appear to be different glycoproteins, and Biomer® and SR desorb a species giving rise to a 1260 cm⁻¹ infrared band. All three polymers lose a protein mixture rich in albumin upon desorption, but Biomer® desorbs the most albumin.

3. The stable layer adhering to the polymer surface shows compositional differences between the three surfaces. SR retains a large amount of glycoprotein compared to Biomer® and PVC. All three surfaces retain a protein mixture rich in fibrinogen, with SR retaining the most fibrinogen and Biomer® the least.

4. On any one polymer surface, there are changes in protein composition with time of adsorption or desorption. This seems to relate to protein adsorption behavior differing on a polymer surface as compared to a protein surface. Thus, adsorbed proteins bound or near the polymer surface differ from adsorbed protein bound or near other proteins.

5. There is replacement of one type of protein with another type of protein during the adsorption and desorption process. Biomer® in particular has been noticed

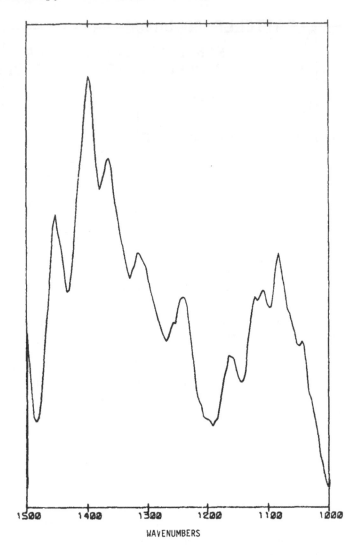

WAVENUMBERS

FIGURE 20.   Spectrum illustrating the appearance of a band at 1361
cm⁻¹ during very early phases of blood-surface contact. This spectrum
was collected 1 sec after whole blood was introduced into the flow
chamber. The relative intensity of this band decreased very rapidly
after this point.

to experience decreases in carbohydrate functional group content (decrease in
glycoprotein) during adsorption 10 to 15 min after the initial blood contact phase.
This is illustrated in Figure 21.

IV. CONSIDERATIONS FOR EX VIVO STUDIES

A. Introduction

The strengths and weaknesses of the FT-IR approach to the study of biological systems should be apparent to the reader at this point. FT-IR is capable of working with
intact physiological systems in an aqueous environment; it works with small amounts
of materials into the hundreds-of-nanograms region. The technique is nondestructive
and allows study of molecular systems without disrupting them in any way. There are
versatile sampling alternatives for studies of bulk, for studies of surfaces, or combi-

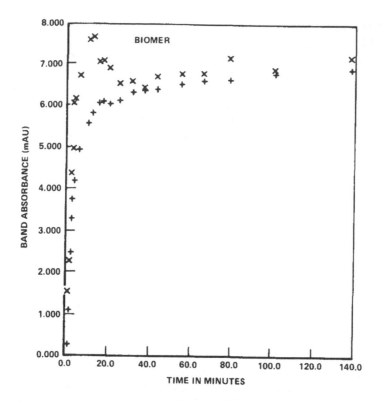

FIGURE 21.   Plot of 1400 cm⁻¹ band and 1080 cm⁻¹ band, normalized
to roughly the same intensity. These data were collected during whole
blood adsorption to Biomer® and indicate that the 1080 cm⁻¹ band (related to glycoprotein carbohydrate functional groups) reached a maximum after approximately 10 min. It then decreased, even though total
protein content continued to increase as shown by the 1400 cm⁻¹ band
intensity.

nations thereof, and it is relatively rapid with good reproducibility. A further advantage of FT-IR as an analytical technique is that it can record near-continuous data,
and it can present those results in real time with a suitably equipped data system. The
importance of real-time data should not be underestimated when undertaking complex
biological experiments. In the course of many of the whole-blood studies that we have
undertaken, the feedback that was obtained saved countless hours of work. The general utility of IR spectroscopy has proven valuable as it allows study of all aspects of
the system. Information on the biological solution, the buffers, and the additives are
available, and this includes quantitative information and information about conformational changes. Another advantage is the fact that the technique is not biased. Since
FT-IR obtains molecular information about the biological system based on the inherent light absorption properties of the molecule itself, there is no need to label or treat
the proteins or macromolecules to study them. This is particularly useful when studying unknown systems, as the technique does not require advance knowledge about the
system, and therefore about what should be labeled. The lack of handling of the sample in contrast to, say, radioiodination studies, where the process of isolating a molecule, radiotagging it, and then restoring it to the system raises many questions about
the validity of the study, is an advantage as well. In the future, we expect to see FT-IR
applied in vivo using infrared fiber optics to deliver and receive the light from embed-
ded internal reflection elements, allowing direct probes of biological systems such as
living animals.

Many of the advantages of FT-IR also cause problems. It is not surprising that a technique which does not require specific labeling raises questions about interpretation and identification of spectra. For example, in the case of whole-blood proteins adsorbing to a surface, all possible proteins can and will be detected by the spectrometer if they are adsorbed. While this has the above-mentioned advantage of not discriminating against unknown species, it makes positive identification of samples very difficult and requires that there be a reference spectrum of each particular species to make positive identifications. The reader should not be discouraged by the fact that there are very few absolutes in biological IR; rather, it is a very versatile technique, the results of which must always be interpreted in light of an overall understanding of the system.

B. Kinetics

When studying protein interactions with surfaces, one would like to learn about the rate of deposition of proteins onto the surface, the relative ratios of proteins on that surface, and the rates of attachment, exchange, and desorption/displacement in a flowing system. FT-IR appears extremely well suited for this type of study. Current spectrometers have the capability of providing sub-1-sec measurement times with reasonable SNR ratios, and this has made it possible to follow protein adsorption to surfaces in a nearly continuous fashion. Since the sample can be monitored in real time, there is no need to stop the experiment to make measurements or perturb the system in any way. Figure 22 shows an example of data collected when plasma is introduced into the flow cell and allowed to interact with the surface. The curve in Figure 22 represents the baseline-corrected infrared band intensity of the Amide II band (1550 cm^{-1}) plotted as a function of time. A valving system is used to control the introduction of plasma into the flow cell. This figure shows a very rapid increase in band intensity primarily due to filling of the cell followed by, or in concert with, rapid adsorption of presumably a monolayer of protein to the bare surface. Adsorption has a somewhat more variable course after this point, generally with an increase in the amount adsorbed onto the surface as a function of time. The rate of increase and the final amount obtained after a sufficiently long adsorption period are determined by both the protein(s) under study and the surface involved. Current work is aimed at further defining the kinetics of the initial contact events between surfaces and protein solutions, but significant improvement over the results shown in Figure 22 will require several instrumental and experimental improvements.

C. Bulk Contribution

In principle, calculation of the amount of protein in the bulk phase which contributes to the infrared spectrum is relatively straightforward. If one defines bulk protein as protein within the evanescent field, but which is at a concentration no greater than that in the initial solution, one may calculate or experimentally determine the effective depth of penetration and therefore the amount of protein that should be observed in the absence of surface adsorption. A second assumption is that there is no water in the adsorbed film. The amount of bulk protein predicted is then based on the amount of water experimentally measured. Since water itself will not adsorb to the surface or become more concentrated, it may be used as an internal standard indicative of the total number of molecules within the evanescent field. At a given concentration, there is a fixed molar ration of protein to water in the bulk solution, and given the infrared water band absorbance, the band absorbance due to solution proteins may be calculated. When an experiment is undertaken, protein adsorption does in fact take place. This results in an increase in the measured amount of protein within the evanescent field, since both the adsorbed and bulk components contribute to the measured spectrum. The difference between the calculated amount (bulk) and the experimentally

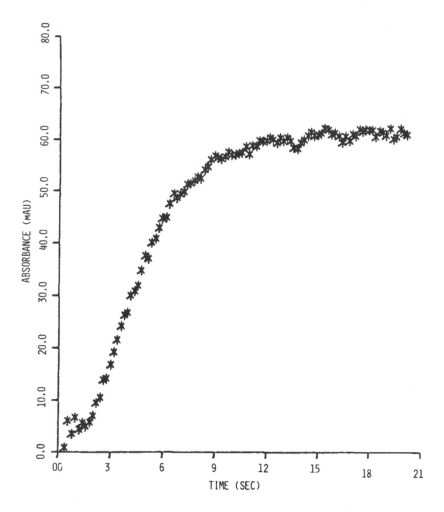

FIGURE 22. High time resolution plot of plasma Amide II band intensity. Ten infrared spectra were collected per second and band intensities measured to obtain this plot.

measured amount (bulk and surface) is the quantity of interest (surface bound). Accurate calculation of the bulk contribution is important for careful study of the initial events of protein interactions with surfaces. Obviously, the more dilute the bulk solution, the less severe the problem. When solution concentrations well below 1 mg/ml are used, it essentially may be ignored. To further complicate the bulk calculation situation, the initial spectra are subject to dilutional effects when the leading edge of the protein solution enters the cell which is already filled with saline. To prevent proteins from being exposed to an air-water interface, the cell must be filled with saline prior to introduction of the protein solution, and there is always some mixing and dilution which occurs during the initial contact of the solution fronts. Studies are now underway to accurately define the filling characteristics of newer-design flow cells, and to more accurately model the exact contributions that occur from proteins in solution as a function of time.

In an attempt to experimentally determine the contribution of bulk protein to spectra obtained from flowing experiments, we undertook a series of experiments in which the starting solution concentration of human serum albumin was varied under constant flow (15 ml/min, ~125 sec⁻¹ shear) conditions. The final adsorbed protein band intensity data are shown in Figure 23 (open dots), and in each case for a range of concentra-

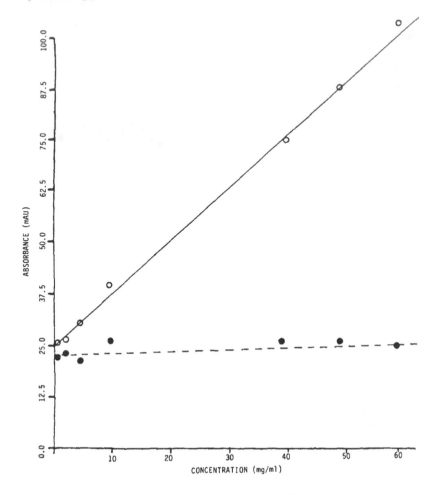

FIGURE 23.   Human serum albumin adsorption. The bulk albumin concentration was varied from 1 to 60 mg/mℓ, and the measured Amide II band intensity in the final spectrum collected during desorption (after 1 hr) is shown (O). Note that the measured intensity is a linear function of concentration (bulk contributions). The intensity of the Amide II band after 1 hr of saline rinse is also shown (•). This implies that a stable adsorb film occurs which is independent of concentration over the range studied.

tions between 1 and 60 mg/mℓ, the final adherent layer after saline desorption (solid dots) was essentially the same. The final adsorbed amount of protein, which is indicated by an Amide II band intensity of 20 to 24 mAU translates to approximately 0.2 µg/cm². Our interpretation has been that albumin adsorbs essentially a monolayer on the germanium surface at each of these concentrations, and no further irreversibly bound protein adsorbs after this point within the concentration range studied.

Thus, for albumin one could say that the amount irreversibly adsorbed is independent of bulk concentration in the range studied. This seems to be a characteristic feature of albumin, and these findings have not generalized to other proteins that have been studied under similar conditions. Since it has been demonstrated that the amount of albumin irreversibly adsorbed to the surface under these conditions is independent of concentration, the measured differences between final band intensities as a function of solution concentration (indicated by the open dots in Figure 23) can be attributed to the bulk contribution.

D. Future Directions

It should be mentioned briefly in this chapter that the interactions which occur between blood and polymer surfaces remain a significant health problem. The adsorption of blood proteins to polymer surfaces, which is often followed by blood coagulation (thrombosis), limits the usefulness of artificial materials as replacement organs. In recognition of this fact, the National Institutes of Health and many biomedical engineering firms have been sponsoring or conducting research on protein-surface interactions, with the intent of defining how different materials interact with proteins and how these interactions might be modified. IR has seemed a promising technique to study this problem area. The ability to study molecular-level events directly, and the ability to study interactions at surfaces using ATR techniques, made IR a logical choice for studies of blood protein-surface interactions.

Our group at Battelle has been applying FT-IR/ATR techniques to protein-surface interaction studies for over 6 years at the time of this writing. The ability to conduct these experiments has developed impressively — flow cell design, experimental set-up, reproducibility, and sensitivity have all undergone great improvement during this time interval. The spectroscopist, unfortunately, has not improved as much as the FT-IR equipment. In the previous chapter the fact was alluded to that there is a significant need for computer-assisted data analysis. What we have learned is that current state-of-the-art equipment is capable of producing much more information than the spectroscopist can digest in any reasonable time period. Thus, either much of the information is simply ignored, or the number of experiments carried out must be severely limited. On the other hand, this flood of information has caused us and other groups working in the biological arena to actively develop computer algorithms for data reduction, summary, and interpretation. Spectroscopically speaking, the most significant progress to date has been the definition of an approach to the problem of interpretation of biological infrared spectra. Considerable effort has been expended determining what type of information is most useful, and strategies to use the computer to assist in analysis and extraction of this information have been formulated.

In summary, studies of protein-surface interactions, including ex vivo studies, are almost routine instrumental experiments. Data interpretation is still extremely difficult and tedious, but the next few years should witness a dramatic improvement. It is certain that much of the molecular-level information that is desperately needed by the bioengineering field can be obtained using FT-IR/ATR. The challenge will be to extract this information from the mass of information which is not relevant to artificially induced thrombosis. As more workers enter the field and develop new areas of application, biomedical FT-IR can be expected to undergo considerable growth. I fully expect that FT-IR will make a number of significant contributions to our basic understanding of life processes.

ACKNOWLEDGMENTS

This work was supported in part by the National Institutes of Health under Grant Nos. HL-24015 and RR-01367.

REFERENCES

1. Gendreau, R. M. and Jakobsen, R. J., Fourier transform infrared techniques for studying complex biological systems, *Appl. Spectrosc.*, 32, 326, 1978.
2. Cameron, D. G. and Dluhy, R. A., FT-IR studies of molecular conformation in biological membranes, in *Spectroscopy in the Biomedical Sciences,* Gendreau, R. M., Ed., CRC Press, Boca Raton, Fla., 1986, chap. 3.
3. Jakobsen, R. J., Brown, L. L., Winters, S., Hutson, T. B., Fink, D. J., and Veis, A., Intermolecular interactions in collagen self-assembly as revealed by Fourier transform infrared spectroscopy, *Science*, 220, 1288, 1983.
4. Cooper, S. L., Application of analytical and spectroscopic techniques to biomedical problems, paper presented at 4th Int. Symp. New Spectroscopic Methods for Biomedical Research, Columbus, Ohio, September 24, 1984.
5. Kellner, R. and Unger, F., *Z. Anal. Chem.*, 283, 349, 1977.
6. Harrick, N. J., *Internal Reflection Spectroscopy,* Interscience, New York, 1967.
7. Fink, D. J. and Gendreau, R. M., Quantitative surface studies of protein adsorption by infrared spectroscopy. I. Correction for bulk concentrations, *Anal. Biochem.*, 139, 140, 1984.
8. Gendreau, R. M., Hutson, T. B., Smith, K. B., and Fink, D. J., Unpublished data.
9. Darst, S. A. and Robertson, C. R., Biological applications of total internal reflection fluorescence, in *Spectroscopy in the Biomedical Sciences,* Gendreau, R. M., Ed., CRC Press, Boca Raton, Fla., 1986, chap. 8.
10. Gendreau, R. M., Leininger, R. I., Winters, S., and Jakobsen, R. J., Fourier transform infrared spectroscopy for protein-surface studies, in *Biomaterials: Interfacial Phenomena and Applications,* Cooper, S. L. and Peppas, N. A., Eds., ACS Advances in Chemistry Series 199, American Chemical Society, Washington, D.C., 1982, 371.
11. Richardson, J. S., The anatomy and taxonomy of protein structure, *Adv. Protein Chem.*, 34, 167, 1981.
12. Richards, F. M. and Wycott, H. W., *The Enzymes,* Boyer, P., Ed., Academic Press, New York, 1971, 647.
13. Susi, H. and Byler, D. M., *Biochem. Biophys. Res. Commun.*, 115, 391, 1983.
14. Walton, A. B. and Blackwell, J., *Biopolymers,* Academic Press, New York, 1973.
15. Woodward, C. K. and Rosenberg, A. J., *J. Biol. Chem.*, 246, 4105, 1971.
16. Shechter, E. and Blout, E. R., *Proc. Natl. Acad. Sci. U.S.A.*, 51, 665, 1964.
17. Iwamoto, R. and Ohta, K., *Appl. Spectrosc.*, 38, 359, 1984.

Chapter 3

FT-IR STUDIES OF MOLECULAR CONFORMATION IN BIOLOGICAL MEMBRANES

David G. Cameron and Richard A. Dluhy

TABLE OF CONTENTS

I. INTRODUCTION

Elucidation of the molecular conformation of a biomolecule is an important step in determining the relationship between physical structure and biochemical function. The conformation of any biological macromolecule governs the physical, chemical, and mechanical properties of the molecule either individually or collectively, as aggregates. In addition, the elucidation of conformational structure for biological macromolecules invariably leads to insights concerning the function of the molecule.

There has been considerable success with the use of spectroscopic methods to determine the conformation of biomolecules. Techniques such as X-ray diffraction, nuclear magnetic resonance (NMR), electron spin resonance (ESR), fluorescence, infrared, and Raman have been used to obtain detailed information about biomolecular structure.[1]

Within the field of membrane biophysics, infrared spectroscopy (IR) in particular offers several advantages for the study of molecular conformation. Infrared absorptions are sensitive to displacements of the dipole moment which in turn depend upon changes in conformation and configuration of the bonds making up the normal mode of vibration.[2] Identification of a group frequency characteristic of a certain aspect of molecular conformation allows the observation of changes in conformation as environmental conditions are varied without the use of perturbing probe molecules. The application of infrared techniques to aqueous biophysical systems has been hindered mainly by the large infrared absorptions of water. To overcome this handicap, several techniques have been developed that allow the use of water as an infrared solvent and are described below.

The current review describes the use of FT-IR to monitor the molecular conformation of model membranes. Unlike other spectroscopic methods, which are restricted to monitoring certain regions of the phospholipid molecule (e.g., ^{31}P-NMR), or rely on the synthetic incorporation of an isotopic probe (e.g., ^{2}H-NMR and ^{19}F-NMR) or introduction of a possibly perturbing probe molecule (e.g., ESR and fluorescence), vibrational spectroscopy has the advantage of noninvasively monitoring absorptions due to the hydrocarbon, interfacial, and head-group regions of the molecule. In addition, the time scale of the infrared experiments (approximately 10^{-12} essentially freezes most molecular motions; therefore, the interpretation of experimental results is not complicated by time-scale averaging of anisotropic motion. Until recently, the principal use of IR in the biological sciences was as an analytical tool for identification and characterization of compounds. It was not, however, considered to be a viable tool for the study of the structure and dynamics of complex biochemical systems.

Three factors can be identified which hindered the use of IR in biochemistry. First, the most relevant solvent for such studies is water. Water is an intense infrared absorber. In the noncomputerized double-beam dispersive instruments in general use until the late 1970s, a pair of cells with the absorption in the reference beam matching that of the water in the sample beam was required. Due to the limited dynamic range of such instruments, short-pathlength (<25 μm) cells were employed; filling, matching, and maintaining such cells was a nontrivial exercise, particularly as the concentrations employed were such that solute would displace a large proportion of the water from the sample cell.

Second, the window materials and the sampling techniques were limited. The most commonly employed infrared windows, NaCl and KBr, are soluble in water. The only windows suitable for general use in transmission spectroscopy were, and still are, CaF_2 and BaF_2. High-quality, high-refractive index KRS-5 and Ge were available; however, they were not suitable for use in transmission spectroscopy. For these materials, the mismatch between the refractive indexes of the windows and the samples results in the superposition of fringes or ''channel spectra'' on the sample; in addition, KRS-5 is etched by water. These materials could be employed as attenuated total reflectance crystals. However, this is a low-throughput technique, necessitating the use of beam condensers and, in double-beam instruments, matched cells, and was even more complex than transmission.

The third factor that delayed the use of IR in biochemistry was the lack of digitization of the data. Most biochemical systems of spectroscopic interest result in small spectral changes in bands that are usually broad and overlapped. The study of such spectra requires highly reproducible spectrometers and the ability to easily process large numbers of spectra using precise algorithms directed at measuring subtle variations.

Recent developments in spectrometers and window materials have greatly facilitated such studies. The greatest advance resulted from the development of FT-IR spectrometers for routine laboratory use. In order to operate, these instruments required a moderate-sized computer with considerable storage capability; this permitted the facile development of data-processing techniques and the routine collection of large numbers of digitized spectra. Extremely high signal-to-noise ratio (SNR) spectra can be collected relatively rapidly (<1 sec per spectrum). In the devising of experiments this can be utilized in two general ways. On one hand, one can collect for a long period in order to study very weak signals. An excellent example of this is the field of infrared vibrational circular dichroism, where signals on the order of 10^{-4} to 10^{-5} absorbance units are measured.[3-5] On the other hand, one may collect for a short period of time (0.1 to 60 sec) to obtain moderate SNRs (1000 to 2000), the advantage then being that many spectra under slightly varying conditions can be rapidly collected. An example where this approach is being pushed to the limit is the in vivo studies of blood-surface interactions (Chapter 2 of this volume, R. M. Gendreau's chapter on applications of biomedical FT-IR to proteins). Within this general context of improved SNR, FT-IR has greatly expanded the range of sampling accessories that can be routinely employed. In particular, attenuated total reflectance (ATR) spectra of extremely high quality can be obtained rapidly, and new accessories such as diffuse reflectance infrared Fourier transform (DRIFT) and infrared microsopes can be employed. In this regard there have been considerable advances in materials. High-quality ZnSe and CdTe ATR crystals are now available in a variety of shapes, some specifically designed to improve the throughput by matching the circular apertures of the FT-IR instrumentation.

In concluding this introductory section we mention two other advantages of FT-IR spectrometers. First, while the fact that high SNRs can readily be achieved is well known, the precision of the frequency scale is not so widely recognized. The reproducibility of the frequency scale is determined by the stability of the wavelength of the light from the He-Ne reference laser. Typically, frequencies are reproducible to within a few thousandths of a wavenumber, regardless of the resolution. This refers only to the reproducibility of the frequency scale. Repeated measurements of the position of a band in a spectrum may give reproducibility close to this, provided the SNR of the absorbance or reflectance scale is sufficiently high. However, variations in sample position, temperature, or changing cells will probably shift the band due to intrinsic shifts (temperature, pressure, etc.) or optical effects (cell shift, cell change, etc.), Second, the instruments are extremely versatile. Spectral ranges from the near UV to the far infrared can be selected by changing the source, detector, and beam splitter. Except in emission experiments, the sample is located after the interferometer and consequently, the detector can be moved and the beam switched to one or more experiments. This removes the constraint that the experiment fit in the sample compartment, greatly facilitating the development of complex experimental apparatus.

II. PRACTICAL CONSIDERATIONS FOR BIOLOGICAL IR

A. Sample Preparation

Difficulties associated with preparing samples for transmission experiments fall into two broad categories: cells and instruments. The cell windows or elements must be kept meticulously clean, particularly if weak bands are to be studied. Finger grease is easily applied and not removed by normal washing. As its spectrum comprises the same bands as those in lipids, its presence effectively negates the study of slight changes in the lipid spectra. We have found that removal requires sonication in a chloroform-methanol mixture, followed by handling of the windows by the edges with protection such as lint-free cotton gloves or disposable plastic gloves (worn once only).

In transmission spectroscopy, the windows must not only be clean, but they must also be polished and flat. The last point is perhaps the most important. Badly pitted plates can be polished quite well. However, the pitting gives rise to a range of pathlengths within the cell. This can result in total absorption at some points in the sample compartment image and partial absorption at other points. The result can be major line-shape distortion, particularly in the region of strong bands.

Given good windows, the most common problem encountered in transmission spectroscopy is the occurrence of fringes of "channel spectra". Even if CaF_2 or BaF_2 windows are used, weak fringes will occur as a result of the difference between refractive indexes of the cell windows (~ 1.5) and the sample ($nH_2O \sim 1.33$). However, this will only be a problem when the sample is clear. Lipid samples are often semiopaque and scatter a large amount of light (evidenced by sloping baselines). Fringing is not a problem under such conditions.

The problems of pitting and fringing are not found in ATR spectroscopy. However, with many cells the empty cell spectrum will include bands from the O-rings or gaskets used to seal the cell. It can be extremely difficult to completely ratio out such bands, due to positional and temperature differences between the sample and reference spectra. Consequently, it can be difficult to measure small changes in sample bands in such spectral regions.

The principal instrumental difficulties are associated with detector nonlinearity and gain ranging, and are discussed elsewhere in this book (Chapter 1, R. M. Gendreau's chapter on equipment and experimental considerations relating to biomedical IR spectroscopy). However, it is worth noting that gain-ranging errors can be accentuated in highly absorbing samples, which give rise to low-dynamic-range interferograms. The consequence can be a fringe running through either a transmission or an ATR spectrum. Such effects are easily checked by running with no gain ranging (the recommended method, in any case) or by changing the gain-range point by a factor of 2. This will halve or double the frequency of the fringes.

Finally, as in dispersive measurements, the optimum peak absorbance in FT-IR is ~ 0.7. Measurements at peak absorbances greater than 1.0 become increasingly prone to noise (due to the log conversion from percent transmission to absorbance) and nonlinearity effects.

B. Data Processing

1. Effect of H_2O

In general, the spectra of biological materials are complex, and the spectral changes observed as a result of a perturbation to a system are small. Consequently, it is often necessary to employ quite complex data-processing techniques. Fortunately, with the large storage facilities of an FT-IR, and the relatively sophisticated computers used to collect and transform the data, such computations are easily performed.

The most fundamental data-processing operation in aqueous IR spectroscopy is the subtraction of the water spectrum. The software makes this a facile operation, provided the water spectra match. There are two main requirements. Cell windows should be of the same material, and pathlengths should be approximately the same (±30%). Given this, it is then critical that the temperatures at which the spectra were obtained be identical. The degree of hydrogen bonding is sensitive to change in temperature and consequently the spectrum is extremely sensitive to temperature variations (Figure 1). We found it necessary to collect a series of reference water spectra, 4°C apart in temperature, and write a program to interpolate the spectra to obtain intermediate temperature spectra. Only when this procedure was used were we able to monitor extremely small changes in frequency and, in particular, bandwidth in bands that were superimposed on the water bands.

As a final point concerning water subtraction, in many systems the lipid concentration will be quite high. In such systems much of the water will be interacting with the lipid head group. This will alter the water spectrum and it will be impossible to completely subtract it out.

2. Frequency and Bandwidth Measurements

By virtue of their mode of operation, FT-IR spectrometers exhibit extremely high reproducibility of the wavenumber scale, regardless of the resolution. This permits the measurement of extremely small frequency shifts, provided an appropriate algorithm is used. Two general classes of algorithm are employed — least-squares and center-of-gravity.[6,7] There is no significant difference between the precision of these methods, provided that they have been programmed correctly.

There is scope for misuse of such algorithms. The algorithms always use data from a number of frequencies to calculate the position of a band. For example, a five-point least-squares algorithm uses a 4 cm^{-1} region of the peak to calculate the position if the spectrum is digitized every 1 cm^{-1}. If the peak has a full-width at half-height (FWHH) of 6 cm^{-1}, this represents a large fraction of the peak; if it has a FWHH of 40 cm^{-1}, a 4 cm^{-1} region may be just noise at the top of the band, and may not contain precise enough data to calculate the position.

In summary, to obtain a precise peak position, the user must ensure that the algorithm uses an appropriate number of data points relative to the width at the band and the encoding interval employed. As the bandwidths within a given spectrum may easily vary by a factor of 10, it is often necessary to employ the same algorithm twice, once with, say 3 points, and once with 30 points.

Bandwidth measurements employ a relatively simple algorithm which can, however, be almost as precise as the peak position algorithms. When employing this algorithm the user is advised to avoid only measuring the FWHH. The algorithm can be written such that the width can be measured at any fraction of the peak height. This is a highly recommended procedure, particularly when studying phase transitions. In many systems the transition is not very cooperative. As a consequence, at least two principal bands will be present during the transition, one from each phase, and their relative intensities will change as the transition progresses. In many cases the overlap of the bands is such that the FWHH values may exhibit a sinusoidal curve during the transition, while the width at nine tenths peak height reaches a maximum just before the completion of the transition, and then decreases.[8,9] This observation of such a plot is, of course, evidence for a noncooperative transition.

Finally, if the user is concerned about the absolute frequency and bandwidth value, the baseline must be taken into consideration. In the case of peak positions, a sloping baseline will shift the peak position toward the frequency at which the baseline has the higher value. Frequency shifts measured on such baselines will be accurate, but absolute values will be incorrect. Generally, if precise comparisons are to be employed we recommend the use of derivatives or deconvolution.

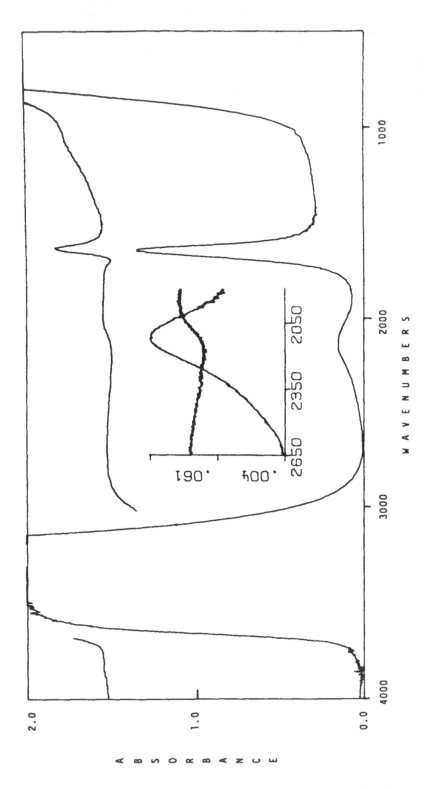

FIGURE 1.   FT-IR absorbance spectrum of doubly distilled water in a 6-μm-thick BaF₂ cell at 33°C (lower curve) and FT-IR difference spectrum (upper curve) obtained by subtracting the above spectrum from one recorded under identical conditions at 43°C. Insert: vertically expanded water and difference spectra of the association band in the region interfering with the C–D stretching modes. (From Cameron, D. G. et al., *J. Biochem. Biophys. Meth.*, 1, 21, 1979. With permission.)

In the case of bandwidth measurements the user should know the way in which the algorithm works. In some cases a sloping baseline is subtracted from the spectrum before the width is calculated. This is the recommended procedure. However, it is possible that the algorithm calculates a horizontal width relative to a sloping baseline. This width will, of course, be highly dependent on the slope of the baseline.

3. Difference Plots

Generally, the difference algorithm is used to subtract the spectrum of one component from a complete spectrum. However, in studies where the spectrum of a system has been recorded as a function of pressure, concentration, time, temperature, etc., in a fixed-pathlength cell it can be used in a different mode. If one has such a series of normalized spectra, one can generate a series of difference spectra, A, of the form

$$A = (B_{n+j} - Bn)/(P_{n+j} - P_n)$$

where B_n is the nth spectrum, P_n is the value of the ranging parameter, and j is an integer. As an example, Figure 2 shows plots of the frequencies of the C=O stretching and CH2 scissoring bands of DPPC as a function of time when the sample was held at 2°C.[10] It is apparent that the bands are changing with time, but little information about the rate of change from such plots is known. Figure 3 shows the difference plots from these bands. They illustrate that there is an abrupt change in the first 9 hr, followed by a gradual increase in the rate of change up to the period 42 to 51 hr. Following this the rate of change decreases. This illustrates that difference plots can yield information on the rate of change of a spectrum that is not available from frequency and bandwidth plots.

4. Deconvolution and Derivation

a. Deconvolution

The spectra of biological samples frequently contain broad, overlapped bands. This leads to uncertainty in band assignments and difficulty in ascertaining what is giving rise to changes in band contours. When such problems are encountered the technique of Fourier deconvolution can frequently be of value.[11-15]

Any recorded spectrum can be expressed as the convolution of a known line shape with a spectrum, i.e.,

$$E = A*B$$

where E is the experimental spectrum, A is the known lineshape function, and B is the other spectrum.

The reader is quite familiar with one example of this. That is the case where A is the instrumental lineshape function (ILSF), and B is the true, noninstrumentally distorted spectrum. The effects of such a convolution are well understood when A is a symmetric function as, to a very high approximation, is the case for grating and FT-IR spectrometers. The spectral lines of E are symmetrically broadened to a degree dependent on the relative widths of A and the lines in B. However, the shape of a given line in E is uniquely determined by its shape in B. Thus, an isolated line which is symmetric in B will be symmetric in E, and a line which is asymmetric in B will be asymmetric in E. A third case is also found; i.e., when lines are partially or fully resolved in B. Convolution of these lines with A may result in partial or complete loss of resolution, giving a complex band contour. However, the contour still represents the sum of the individual lines broadened by A. Exactly the same criteria apply when we consider A to be a purely mathematical symmetric function.

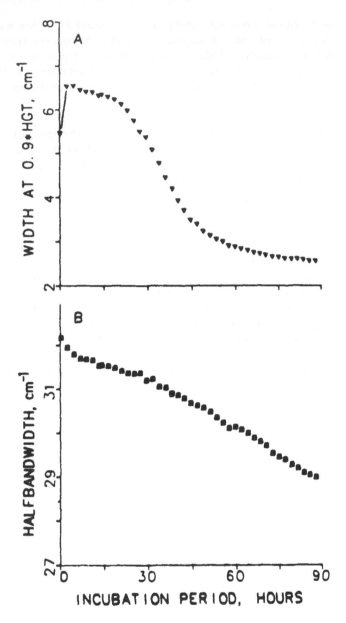

FIGURE 2.   Dependence of the bandwidths of the CH_2 scissoring and C=O stretching bands of DPPC on the period of incubation at 2°C. (A) Full-width at nine tenth peak height (0.9 * HGT) of the CH_2 scissoring band; (B) full-width at half-height (0.5 * HGT) of the C=O stretching band. (From Cameron, D. and Mantsch, H., *Biophys. J.* 38, 175, 1982. With permission.)

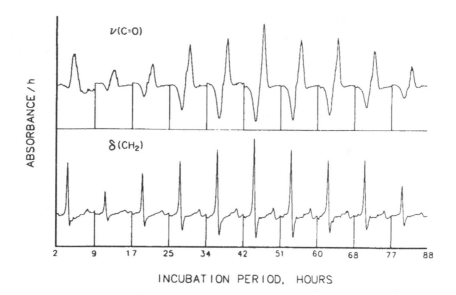

FIGURE 3.   Infrared difference spectra in the regions of the C=O stretching (top) and CH₂ scissoring (bottom) bands during the incubation at 2°C. Spectra have been normalized with respect to the time interval used to generate the spectra. The times used in the subtraction are indicated on the bottom axis; e.g., the left-most difference spectra were generated by subtracting the spectrum recorded after 2 hr from that recorded after 9 hr and dividing the result by 7. (From Cameron, D. and Mantsch, H., *Biophys. J.*, 38, 175, 1982. With permission.)

The above results are, of course, only observed when the ILSF is similar in width to or broader than the lines being studied. However, with biological samples it is usually easy to run the spectrum at a resolution high enough that the distortion of B is minimal, i.e.,

$$E \sim B$$

As a rough guide, this will be achieved when the ILSF is at least two times narrower than the lines of interest.

In this situation we can use mathematical deconvolution, choosing a lineshape A and solving Equation 1 for B. Just as a high-resolution spectrum will, when convolved with a symmetric ILSF, give a low-resolution spectrum, if we take the deconvolved spectrum and convolve it with A, we will recover our spectrum. Continuing the analogy, the number of lines in B and their positions, integrated intensities, and general shapes (symmetric, asymmetric, etc.) will be the same as they were in E. However, as is the case when spectra are recorded at different resolutions, the relative peak heights will differ and, if a contour was composed of several overlapped lines, they will be resolved to some degree.

The procedures used are well known and employ Fourier transforms. Details can be found in References 11 and 14. When employing the procedures, two principal parameters must be optimized, the width of A and a parameter, K. This latter parameter is the ratio of the width of a line before deconvolution to its width after deconvolution. Thus, if K = 3, a line with a width of 12 cm⁻¹ in E would have a width of 4 cm⁻¹ in B.

The first step is to isolate a section of the spectrum where the lines have similar widths; e.g., the C—H stretching or the C=O stretching region. K is then set to a low value (2 is usually appropriate) and an estimate of the width of A is made. This value should approximate the width of the lines in the region being deconvolved (e.g., 10 to

20 cm^{-1} for C=O stretching bands). The result is then displayed. If the width is too great, negative lobes will be observed and the width should be reduced; if lobes are not present, the width is then held constant and K is increased. When the noise increases to an unacceptable level, the appropriate value for K will be slightly lower.

Some programs include one other parameter which can be optimized, the lineshape. As is the case with instruments, the lineshape is always symmetric. The best results are obtained when the lineshape approximates the actual spectra lineshapes. In IR spectroscopy most lineshapes are generally close to Lorentzian, hence a Lorentzian lineshape is used. However, in some cases, particularly when hydrogen bonding is occurring, the lineshapes may be somewhat Gaussian. It is then advantageous, after selecting the width and K, to introduce a Gaussian component into the lineshape and then, possibly, increase K.

As a final point, note that this is not an exercise in curve fitting and hence there is no attempt made to match the lineshape exactly. All that is necessary is to approximate it well enough that the result shows resolution enhancement without spectral artifacts.

b. Derivation

Mathematically, even derivatives are closely related to deconvolved spectra.[13,15] Both enhance resolution at the expense of the SNR, and both are prone to the same artifacts. We recommend the use of derivatives as a supplement to deconvolution. Generally, the available software is not as sophisticated as the deconvolution software. However, with care satisfactory results can be achieved.

First, select a region of the spectrum in which the bands have similar widths. It is a common misconception that derivatives of broad bands are not useful. This is only so when narrow lines are included in the region being differentiated.

Next, smooth the region and subtract it from the original spectrum. Continue doing this until the difference spectrum shows spectral features as well as noise. Decrease the smoothing slightly. This will only work if the smoothing is least-squares; Lorentzian smoothing will always produce spectral features and the procedure will not work as well.

Take the second and fourth derivatives and compare them. If the second derivative is noisy, discard the fourth derivative and smooth the second derivative until the noise level is acceptable. If the second derivative is noise free and the fourth is noisy, smooth the fourth derivative. You may achieve a result in which the derivative features are still sharper than they are in the second derivative.

c. Problems in Deconvolution and Derivation

Both deconvolution and derivation can readily produce artifacts. However, with care, they can equally well produce valuable insights into the nature of spectra. The most commonly encountered artifacts are outlined below.

First, both trade off SNR ratio for information. Consequently, the spectrum must have a high SNR to start with. Note also that the noise in a deconvolved or differentiated spectrum is no longer random but is highly periodic. The user must be careful that such noise is not confused with peaks.

Second, they can accentuate weak features in the spectrum. In particular, water vapor lines can be enhanced, as can fringes. The user must ensure that the spectrum is free from such features.

Third, nonlinear effects can lead to artifacts. Nonlinearities in both the peaks and the noise are greatest at absorbances greater than 1, so peak heights should be kept below this.

Fourth, ringing artifacts may occur at the edges of the deconvolved region. This is most frequently observed when the region selected terminates part way up the side of an adjacent band. As far as possible, the region selected should terminate at a point of low gradient.

Finally, a highly recommended procedure is to use a peak-generating program to generate a synthetic spectrum with bands of different widths, heights, and degrees of overlap. Deconvolution and derivation should then be performed with and without the presence of noise. As the answer is known, it is easy to become familiar with the procedures and to recognize when they are being misused.

5. Least-Squares Curve-Fitting

Least-squares curve-fitting covers the general class of techniques whereby one attempts to minimize the sum of the squares of the difference between an experimental spectrum and a computed spectrum generated by summing several curves.[16-18] The curves may be actual, spectral (used in quantitative analysis), or synthetic (Lorentzian, Gaussian, etc.). The programs will vary in sophistication, from those that allow parameters to be held fixed, to those in which all parameters are varied in the optimization procedure.

The general advice for these programs is that they be used with extreme caution. This is particularly so when fitting several synthetic curves to a complex contour. First, infrared lineshapes are seldom, if ever, completely symmetric. Second, it is always possible to fit additional synthetic curves to the contour. The consequence of this is that it is generally possible to obtain a variety of solutions to a given problem. The best approach is to use the smallest number of curves that one can demonstrate are present by means of deconvolution and derivation.

III. FT-IR STUDIES OF MODEL AND NATURAL MEMBRANES

A. Characteristic IR Spectral Regions for Model Biomembrane Studies

1. C−H Stretching Region

The thermotropic polymorphism of simple model phospholipids (e.g., 1,2-dipalmitoyl-sn-glycero-3-phosphocholine or DPPC) has been extensively studied; all the phospholipid model membranes under consideration here form bilayer structures when hydrated and exist in two distinct phases. At low temperatures in the gel phase below its characteristic phase transition temperature (T_m), the acyl chains of phospholipids exist in a regular lattice structure (see below) characterized by an extended, mostly all-trans hydrocarbon chain. As the temperature increases and approaches T_m, gauche conformers are introduced into the rigid lattice and the regular packing is disrupted. Above T_m in the liquid-crystalline phase, large numbers of gauche conformers occur, resulting in a fluid acyl chain structure (see Reference 19 and references contained therein).

As the gel-to-liquid-crystalline phase transition of phospholipids involves primarily the introduction of conformational disorder into the lipid hydrocarbon chains, the region of the infrared spectrum between 3100 to 2800 cm^{-1} that contains the carbon-hydrogen stretching fundamental vibrations has been extensively used to characterize the lipid phase state. In the case of phospholipids, the infrared spectroscopic parameters used to identify phase changes include band frequency, halfwidth, peak height, and integrated intensity.[20] Figure 4 illustrates the temperature dependence of the conformation-sensitive C−H stretching region for DPPC. The spectral features arise from the asymmetric and symmetric CH$_3$ stretching modes near 2956 and 2872 cm^{-1}, while the antisymmetric and symmetric CH$_2$ stretching bands are observed at 2920 and 2850 cm^{-1} in addition to a broad Fermi resonance band centered at about 2900 cm^{-1}.[21]

An increase in temperature results in a reduction of peak height and intensity, an increase in half-width, and a shift in frequency to higher wavenumbers, with all the parameters showing maximal discontinuities at the transition temperature.

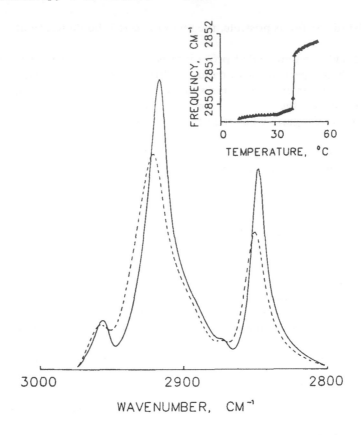

FIGURE 4.   Temperature dependence of the infrared spectra of 1,2-dipalmitoyl-sn-glycero-3-phosphocholine (DPPC) in the region of the C—H stretching bands. The upper (solid) spectrum represents DPPC in the gel phase at 10°C. The lower(dashed) spectrum represents DPPC in the liquid-crystalline phase at 50°C. Insert: the relationship between frequency and temperature for the CH_2 symmetric stretching band of DPPC.

As mentioned above, the temperature-dependent spectral changes can be attributed on a molecular level to variations in the conformation of the mostly all-trans phospholipid acyl chains in the gel phase, with the introduction of a large number of gauche conformers at the characteristc phase-transition temperature.[19] As has been previously demonstrated and shown in Figure 4, the frequencies of the CH_2 stretching bands are sensitive indicators of the phospholipid-gel-to-liquid-crystalline phase transition.[22] The sensitivity of the variation of C—H frequency to phase changes for various hydrocarbon systems can be traced to changes in the interaction constants between adjacent CH_2 groups as the relative orientation of the groups change with the introduction of gauche conformers.[21]

As an empirical measurement correlated with the conformation of the lipid acyl chains, the frequency of the antisymmetric CH_2 stretching band at 2920 cm⁻¹ or the symmetric stretching band at 2850 cm⁻¹ has been extensively used to monitor the effects of ions, cholesterol, and membrane proteins on lipid conformation.[9,23-26] Pure, saturated, synthetic phosphatidylcholine (PC) model membranes have been the most extensively studied of the compounds used as membrane models. The thermotropic behavior of these phospholipids has been examined by FT-IR.[27] The detailed temperature profiles of the frequency of the 2920 or 2850 cm⁻¹ infrared bands show discontinuities that correlate with the corresponding calorimetric phase transition for the particular lecithin under study; hence, these parameters can monitor the average conformational order of the system.

The second largest class of membrane phospholipids after the ubiquitous phosphatidylcholines is the phosphatidylethanolamines (PE). PEs containing saturated acyl chains exhibit phase transitions at higher temperatures than the corresponding PCs.[28] The difference can be accounted for by considering the interaction of the corresponding head groups with solvent (H$_2$O). A recent X-ray structure of a saturated PC has suggested that the phosphate group in these molecules is linked by intervening water molecules.[29] In the smaller PE head group, however, it is the ethanolamine moiety that is linked to the adjacent phosphate.[30] The result is that PEs are difficult to hydrate and exhibit a phase transition at a higher temperature than the corresponding PCs.

FT-IR spectra of DPPC, DPPE, and the partially N-methylated DPPE derivatives in the C—H stretching region suggest that more conformational disorder is introduced at T$_m$ in the PE compounds than in the PC and that the amount of disorder can be inversely related to the degree of methylation of the head group amine.[31] This conclusion is derived from the fact that the magnitude of the frequency shift of the 2850 cm^{-1} band at the T$_m$ of the lipid is related to the amount of conformational disorder introduced. Experimentally, the magnitude of these shifts is measured as 1.8, 2.1, 2.2, and 4.5 cm^{-1} for DPPC, dimethyl DPPE, monomethyl DPPE, and DPPE, respectively.

While the model phospholipids discussed until now all exist as stable bilayers, other structural arrangements are possible for PEs containing highly unsaturated acyl chains.[32] In particular, the lamellar, liquid-crystalline PE membrane may rearrange under appropriate conditions to a nonlamellar inverted hexagonal (H$_{II}$) phase in which the PE head groups are oriented toward a central cylindrical aqueous channel, while the acyl chains are oriented outward from the central cylinder. This change in orientation, which is due to an increasing concentration of gauche bonds in the acyl chains,[33] can be monitored via the frequency of the C—H stretching vibrations at 2850 cm^{-1}. As an illustration, two transitions are observed in egg-yolk PE by FT-IR.[34] The first shows a frequency shift of approximately 2 cm^{-1} at 12°C; a further frequency shift of about 1 cm^{-1} occurs at 28°C (Figure 5). The first transition is associated with the gel-to-liquid-crystalline phase transition while the second corresponds to the bilayer-to-inverted-hexagonal phase. These types of experiments demonstrate the ability of FT-IR to precisely monitor extremely small spectral changes ($\sim$1 cm^{-1}) and correlate these changes to a structural reorganization in the membrane.

In addition to studies of pure phospholipids, physical studies of lipid-protein complexes using various spectroscopic methods have led to conflicting conclusions concerning the physical conformation of the phospholipid molecules in the immediate vicinity of intrinsic membrane proteins (for a review, see Reference 35). The question of whether an immobilized class of lipid immediately adjacent to the protein exists has also been the subject of some debate.[35] In addition, the topic of lipid-protein interactions has been extensively studied due to the marked dependence of membrane protein function on the physical state and chemical structure of phospholipid in contact with the protein in the membrane.[19]

Several studies have appeared using FT-IR to investigate the above questions. Mendelsohn et al.[25] have reconstituted glycophorin, an integral membrane protein isolated from human erythrocytes, into unilamellar vesicles with 1,2-dimyristoyl-sn-glycero-3-phosphocholine (DMPC) at lipid-to-protein mole ratios ranging from 50:1 to 200:1. As shown in Figure 6, plots of frequency vs. temperature for pure DMPC and the lipid-to-protein complexes indicate that incorporation of protein results in very little change in conformational order, either at low temperatures in the gel phase or at high temperatures in the liquid-crystalline phase. It does, however, progressively broaden the transition as the proportion of protein is increased, thereby decreasing the cooper-

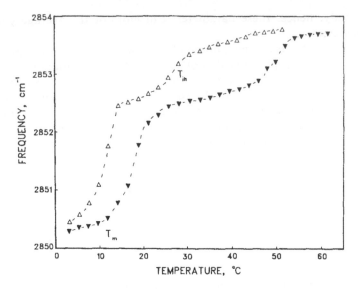

FIGURE 5.   Temperature dependence of the infrared frequency of the symmetric CH_2 stretching vibration in natural egg yolk PE (△) and in PE obtained from egg yolk PC (▼). (From Mantsch, H. H. et al., *Biochemistry*, 20, 3138, 1981. With permission.)

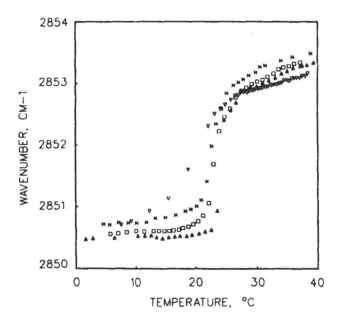

FIGURE 6.   Plots of frequency vs. temperature for the symmetric CH_2 stretching band in the infrared spectra of DMPC multilayers (▲) and DMPC-glycophorin complexes at 200:1 (□), 100:1 (•), and 50:1 (▽). (From Mendelsohn, R. et al., *Biochemistry*, 20, 6699, 1981. With permission.)

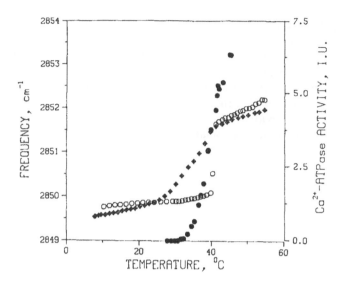

FIGURE 7. Plots of the temperature dependence of the lipid CH₂ symmetric stretching frequency for a DPPC-Ca²⁺-ATPase complex (◈) as well as for pure DPPC multilayers (O). Included for comparison are data for the temperature dependence of ATPase activity (•). (From Mendelsohn, R. et al., *Biochim. Biophys. Acta*, 775, 215, 1984. With permission.)

ativity of the lipid-phase transition. There is no evidence from the data at high temperature of a second melting transition that might be associated with the melting of an immobilized boundary layer.

A second study using glycophorin as the integral membrane protein examined its interaction with the anionic membrane phospholipid phosphatidylserine.[36] In this case, as distinct from the previous DMPC-glycophorin study, the protein nearly abolishes the gel/liquid-crystalline transition seen in pure PS and introduces static disorder into the gel phase of the PS, the result of being unable to pack efficiently into the acyl chain environment. The data suggest a preferential PS-glycophorin interaction.

Although glycophorin has been extensively used as a model for an integral membrane protein, it suffers from the disadvantage of not having a function that can be easily assayed. Thus, experiments designed to investigate the dependence of membrane protein function as a consequence of the physical state or chemical structure of the surrounding phospholipid cannot be performed using this particular protein.

One system that has been widely used in studies of lipid-protein interactions and the relationship of the activity of membrane-bound proteins to lipid structure is muscle sarcoplasmic reticulum vesicles. The major protein of the native membrane is a (Mg²⁺ + Ca²⁺)-ATPase (ATP phosphohydrolase, E. C. 3.6.1.3) which is involved in Ca²⁺ transport.[37] Reconstitution procedures have been developed by which ∼95% of the native lipid may be replaced with external lipid and still retain full enzymatic activity.[38] Thus, the effect of protein on lipid phase behavior may be studied as well as the effect of lipid physical state on enzymatic function.

Recently, an FT-IR study of lipid-protein interactions in reconstituted sarcoplasmic reticulum has appeared.[39] Reconstitution of the Ca²⁺-ATPase component of the membrane with DPPC (lipid to protein mole ratio 30:1) leads to purified vesicles in which the level of DPPC enrichment is measured at 95.8%.

Figure 7 presents plots of the temperature-induced variation in the CH₂ symmetric stretching frequency at 2850 cm⁻¹ for pure DPPC as well as the DPPC:Ca²⁺-ATPase complex. The pure DPPC bilayers melt cooperatively at 41°C. In the DPPC:Ca²⁺-

ATPase system, the onset of the phase transition is shifted to lower temperatures (27 to 28°C) with the completion of melting at 41°C. In analogy with the glycophorin reconstitution experiments described above, the frequency of the CH_2 bands for DPPC in the reconstituted complex is not significantly different from that of pure DPPC at low temperatures in the gel phase or at high temperatures in the liquid-crystalline phase. Instead, the transition is considerably broadened and shifted to lower temperatures. Again, there is no evidence for a second melting event above 41°C in the lipid protein complex which might be associated with a postulated immobilized component.

Figure 7 also has plotted the temperature-dependent enzyme activity of the Ca^{2+}-ATPase on the same temperature scale as the FT-IR frequency parameter. It is easily seen that the temperature dependence of the enzyme activity closely parallels the melting characteristics of the lipid in the lipid protein complex. Significant ATPase activity begins at 29°C, in close agreement with the 27 to 28°C onset melting temperature of the DPPC. These results demonstrate the requirement of a fluid lipid environment for the expression of enzymatic activity in this system.

2. Deuterated Probes

As discussed in the previous section, the importance of the phospholipid phase transition to the structure and function of biological membranes has been extensively investigated in recent years using a variety of spectroscopic probes;[1] such processes as enzyme activity, active transport, and ion permeability have been shown to be dependent on the lipid-phase state.[19,35] In addition, in complex mixtures of lipids, the coexistence of laterally segregated domains of lipids differing in their physical structure may provide a particular chemical environment conducive to functional properties.[40]

The simplest model systems that can exhibit the types of phase behavior and melting characteristics of in vivo systems and yet be amenable to detailed physical investigation are binary lipid mixtures.[41] However, since the majority of biophysical membrane studies have focused on a single lipid species, little information is available on the physical and chemical structure of binary lipid mixtures.

FT-IR spectroscopy offers several advantages for the study of binary mixtures. In particular, if one component of the mixture is isotopically labeled in its conformation-sensitive region, the conformation of each component may be monitored simultaneously. Such an approach has been applied to phospholipids,[42-44] where the synthetic incorporation of perdeuterated acyl chains allows the monitoring of the conformation-sensitive $C-D$ stretching bands in a manner analogous to that of a proteated hydrocarbon (see previous section). Substitution of deuterium for hydrogen results in a frequency shift of $\sim(2)^{1/2}$ for the $C-D$ stretching bands as compared to the $C-H$ stretching bands. As a result, the new $C-D$ stretching bands are located in the spectral range 2300 to 2000 cm^{-1}. In addition, for phospholipid mixtures, the $C-D$ stretching bands of the deuterated chains occur in a spectral region free of interference from other vibrational bands and are easily observed; consequently, the conformation of both components of a binary mixture may be monitored.

As an example of the use of this procedure for studying binary lipid mixtures, Figures 8A and B give the temperature dependence of the $C-H$ and $C-D$ stretching bands for a binary mixture of DPPC and DMPC-d$_{54}$ (DPPC:DMPC-d$_{54}$ mole ratio, 0.75:0.25). In the $C-D$ stretching region (Figure 8B) the CD_3 groups give rise to bands at 2212 (asymmetric stretch) and 2169 cm^{-1} (symmetric stretch) while the bands at 2194 and 2089 cm^{-1} are the antisymmetric and symmetric CD_2 stretching bands, respectively.[45,46]

Figures 8C and D show the temperature-induced variation in the symmetric CD_2 and CH_2 frequencies for both components of the binary mixture. As expected for a binary mixture displaying complete miscibility,[9,47] the frequency changes for the components

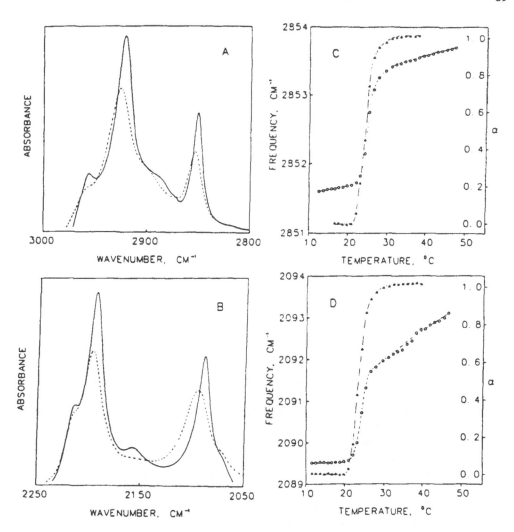

FIGURE 8.   (A) Temperature dependence of the infrared spectra of the C—H stretching region of the DPPC component of a 75:25 binary mixture of DPPC: DMPC-d_{54}. (B) The C—D stretching region of the DMPC-d_{54} component of the same binary mixture as in A. For purposes of display, linear baselines extending from 3000 to 2800 cm^{-1} (in A) and 2300 to 2000 cm^{-2} (in B) have been subtracted. Solid lines indicate low-temperature (15°C) spectra while dashed lines indicate high-temperature (45°C) spectra. (C) Temperature-induced variation in the frequency of the CH$_2$ symmetric stretching band of DPPC (open symbols), along with the α values (closed symbols) calculated according to the procedure described in the text. (D) Temperature-induced variation in the frequency of the CD$_2$ symmetric stretching band of DMPC-d_{54} (open symbols) along with the calculated α values (closed symbols).

of the mixture occur over a wider temperature interval than that of a pure lipid (compare Figures 1 and 8). A further confirmation of a broadened transition in the binary mixtures comes from the calorimetric data for various mole ratios of DPPC and DMPC-d_{54}.[47] Several FT-IR studies of binary phospholipid mixtures and ternary mixtures (that include protein) have appeared that utilize the isotopic substitution approach described above.[9,39,43,47-49]

The binary phospholipid mixture DPPC:DMPC-d_{54} has been characterized by FT-IR[9,47] by assuming a two-state model for the phase transition and calculating a fractional conformational parameter that describes the degree of transition. Calculations based on the Zimm-Bragg theory of cooperative conformational transitions allow ther-

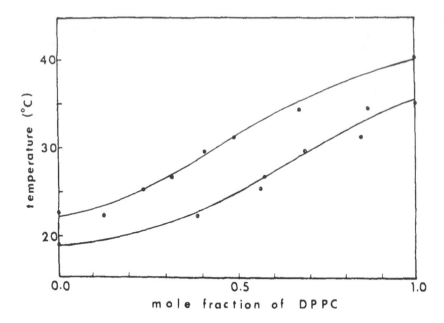

FIGURE 9.   Phase diagram calculated for the binary lipid mixture DPPC:DMPC-d_{54}. The melting temperatures at the end points (0.0 and 1.0 mole fraction of DPPC) were determined from FT-IR melting curves of the pure lipids DMPC-d_{54} and DPPC, respectively. (From Dluhy, R. A. et al., *Biochemistry*, 22, 1170, 1983. With permission.)

modynamic quantities to be derived using the conformational parameter in conjunction with calorimetric measurements.

The application of this model to DPPC:DMPC-d_{54} phospholipid binary mixtures provides evidence for lateral phase separation and the coexistence of well-defined structural domains of phospholipid. In addition, the data suggest preferential clustering of one of the lipid components at the interface between the two phases. These results agree with recent experiments that suggest the compositional segregation of lipids at the phase boundaries of binary lipid vesicles may be responsible for permeability phenomena.[50,51]

The effect of protein on lipid structure and phase behavior has also been studied for DPPC:DMPC-d_{54} vesicles reconstituted with glycophorin.[9] The binary lipid vesicles have been shown by calorimetry to form a nearly ideal mixture.[41,52] By use of the two-state protocol described above, approximate phase diagrams for the binary lipid mixture can be constructed from the FT-IR melting curves. This approach is particularly useful in the ternary system that includes protein since the construction of phase diagrams by traditional means is precluded.[9]

Figures 9 and 10 illustrate the phase diagram for DPPC and DMPC-d_{54} in the absence and presence of protein, respectively. The phase diagram for the binary mixture (Figure 9) is similar to those in the literature constructed from a variety of techniques for the DPPC:DMPC system.[52] Incorporation of glycophorin at a level of about 1 mol% causes a lowering of the liquidus line while leaving the solidus line virtually unchanged (Figures 9 and 10). Such behavior is consistent with the reduction of the enthalpy of melting of the components.[53] The phase diagram for the ternary system (Figure 10) also shows that glycophorin does not induce phase separation in the lipid mixture, as indicated by the absence of monotectic behavior in the solidus line.

The interaction of glycophorin with mixtures of phosphatidylserine (PS) and phosphatidylcholine (PC) was studied in order to observe whether the protein would pref-

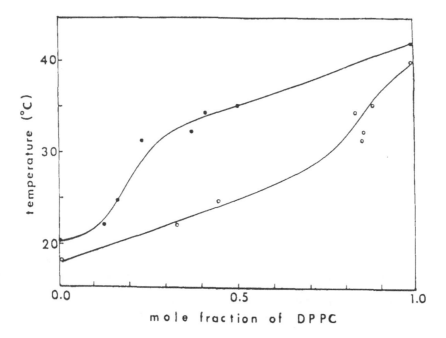

FIGURE 10. Phase diagram calculated for the ternary system DPPC:DMPC-d₅₄: gly-cophorin. The melting temperatures at the end points 0.0 mole fraction DPPC (corresponding to DMPC-d₅₄) and 1.0 mole fraction of DPPC were determined from FT-IR melting curves of a 100:1 DMPC-d₅₄: glycophorin complex and a 100:1 DPPC:glycophorin complex, respectively. (From Dluhy, R.H. et al., *Biochemistry*, 22, 1170, 1983. With permission.)

erentially interact with one component of a binary mixture that differs substantially in physical structure.[49] Figure 11 illustrates the temperature-induced variation in the C—H and C—D stretching bands for the binary mixture PS:DPPC-d₆₂ (30:70 mol %) in the presence and absence of protein. As seen in Figure 11, the bovine brain PS exhibits a broad, noncooperative melting profile in the binary mixture which is accentuated in the presence of protein. In contrast to the PS, the infrared melting curve for the DPPC-d₆₂ component in the presence of protein demonstrates a 6-K increase in melting temperature when compared with the binary mixture. In addition, the onset temperature is significantly raised. The FT-IR data suggest that addition of glycophorin to the binary lipid mixture produces a preferential interaction with the PS component. Consequently, the residual bulk lipid becomes enriched in DPPC-d₆₂ and undergoes a phase transition as observed by FT-IR at a temperature higher than the 30:70 binary lipid mixture.

Analogous experiments have been performed using the Ca^{2+}-ATPase from sarcoplasmic reticulum vesicles reconstituted into DOPC:DPPC-d₆₂ binary lipid mixtures.[39] The effect of protein on the DOPC component is to reduce the frequency of the CH_2 stretching bands at all temperatures, as was observed for the direct interaction of DOPC with Ca^{2+}ATPase. For the DPPC-d₆₂ component, the effect of protein is to increase both the onset and completion melting temperatures. By using an argument similar to that for the PS:DPPC-d₆₂:glycophorin system described above, it was concluded that the Ca^{2+}-ATPase had effectively removed the DOPC from the bulk lipid into DOPC:Ca^{2+}-ATPase domains, thereby resulting in a bulk membrane enriched in DPPC-d₆₂, as seen in the increased melting profile for the DPPC-d₆₂ in the ternary system.

In addition to the use of completely perdeuterated acyl chains, it is possible to syn-

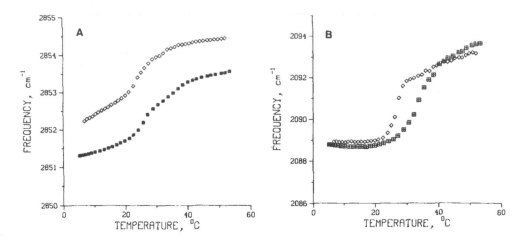

FIGURE 11.   Plots of the temperature-induced variation. (A) The CH_2 symmetric stretching frequency of the PS component in a 30:70 binary combination with DPPC-d_{62} (◇), and in the ternary system PS:DPPC-d_{62}:glycophorin (▥), PS:DPPC-d_{62} mole ratio is 30:70, lipid:protein mole ratio is 160:1. (B) The C—D symmetric stretching frequency of the DPPC-d_{62} component in a 30:70 binary combination with PS (◇) and in the ternary system described above (▥). (From Mendelsohn, R. et al., *Biochim. Biophys. Acta*, 774, 237, 1984. With permission.)

thetically incorporate an isolated CD_2 group at a specific position along the acyl chain of the phospholipid molecule. The advantage of this approach is that the structure of an isolated position in the molecule may be probed. This method has been extensively studied by ^{2}H-NMR;[54] however, until recently, ^{2}H-NMR was restricted to observing the structure of the molecule in the liquid-crystal phase, due to the extreme broadening of the powder pattern in the solid-gel phase.[55] As FT-IR is not restricted to observing only one temperature phase, detailed melting profiles for isolated positions on the acyl chain may, in principle, be obtained.

Cameron, et al.[56,57] have studied the FT-IR spectra of DPPC that have been specifically labeled at the 2, 3, 7 and 8, 13, and 16 positions in the sn-2 palmitoyl chain.* Their findings are summarized in Figure 12. They indicate that throughout the gel phase the acyl chains are extended and contain no substantial numbers of gauche conformers. The main melting transition results in changes in the C—D stretching bands for 2-(3-d_2)DPPC, 2-(7,8-d_4)DPPC, and 2-(13-d_2)DPPC that indicate the introduction of large numbers of gauche bonds. However, the data for 2-(2-d_2)DPPC are complicated by the position of the CD_2 group to the C=O ester which results in large inductive effects. For 2-(16-d_3)DPPC, observation of the progressive splitting in the asymmetric CD_3 stretching band as the temperature is lowered indicates a progressive decrease in the rate of motion of the terminal methyl group in the gel phase.[56]

3. CH₂ Scissoring and Wagging Bands

In addition to the C—H stretching region, two other regions of the mid-infrared spectrum have proven informative in describing the structure of the hydrocarbon chains of phospholipids, especially in the low-temperature gel phase. They are the CH_2 scissoring region between 1480 to 1450 cm^{-1} and the region between 1360 to 1190 cm^{-1} that contains the methylene wagging band progression.[22,58,59]

The detailed organization of the gel phase in pure phospholipids that contains satu-

---

*   The generally accepted nomenclature assigns the number 1 to the carbonyl carbon in the acyl chain with values increasing with the distance from the carbonyl carbon. Thus, the 2 position is directly adjacent to the carbonyl while the 16 position is the terminal methyl in the palmitoyl chain.

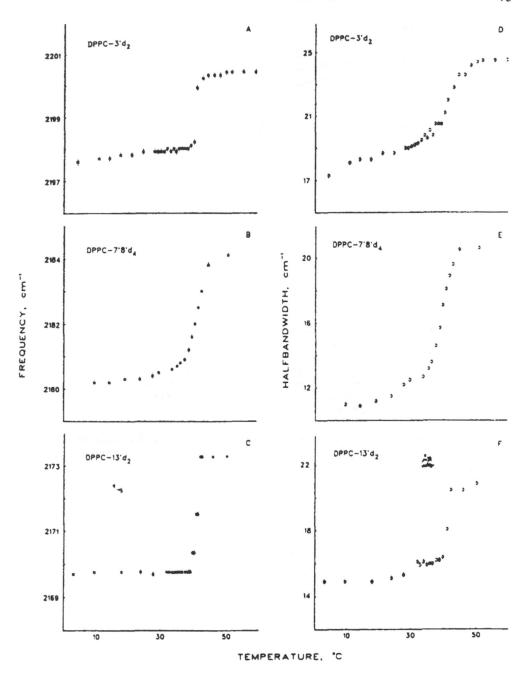

FIGURE 12. Temperature dependence of the frequencies of the antisymmetric CD_2 stretching bands of (A) 2-(3-d_2)DPPC; (B) 2-(7,8-d_4)DPPC; and (C) 2-(13-d_2)DPPC. Temperature dependence of the bandwidth of the antisymmetric CD_2 stretching bands of (D) 2-(3-d_2)DPPC; (E) 2-(7,8-d_4)DPPC; and (F) 2-(13-d_2)DPPC. (From Cameron, D. G. et al., *Biophys. J.*, 35, 1, 1981. With permission.)

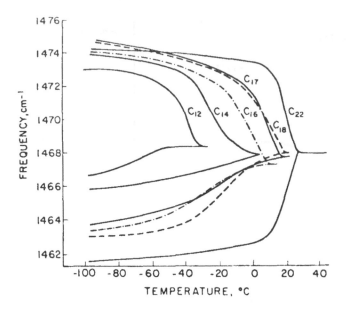

FIGURE 13.   Temperature-dependent factor group splitting of the acyl chain CH_2 scissoring mode of the diacyl phosphatidylcholines. In order to contain the data on one plot and permit a visualization of differences resulting from variations in the chain length, continuous curves have been plotted through the experimental data, and error bars have been omitted. (From Cameron, D. G. et al., *Biochemistry*, 19, 3665, 1980. With permission.)

rated hydrocarbon chains is not as clearly understood as the well-documented liquid-crystalline phase. The gel phase has been shown to be metastable; at 35°C (in the case of DPPC) a second phase transition much reduced in enthalpy from the main transition at 41°C occurs, which is usually referred to as the pretransition.[52] X-ray diffraction studies have indicated that at temperatures near the pretransition (T_p), the acyl chains are packed in a slightly distorted hexagonal form and tilted with respect to the plane of the bilayer. The angle of tilt increases as the temperature is lowered and the packing becomes increasingly nonhexagonal.[60]

IR spectroscopy provides an extremely sensitive method for monitoring the solid-phase behavior of hydrocarbons, including the acyl chains of phospholipids. When hydrocarbons are packed in an orthorhombic crystal lattice common to polyethylene and *n*-alkanes, the resulting interchain coupling of vibrational modes results in a factor group (or correlation field) splitting of the CH_2 scissoring band at 1468 cm^{-1} and the CH_2 rocking band at 720 cm^{-1}.[61,62] Since this factor-group splitting is specific only to orthorhombic subcells, its observation allows the characterization of the packing of the hydrocarbon chains from among the alternative subcells (orthorhombic, hexagonal, or triclinic). In addition, the magnitude of the band separation is indicative of the degree of interchain interaction.[63]

Figure 13 illustrates the temperature-dependent factor-group splitting of the acyl chain CH_2 scissoring mode for a series of diacyl phosphatidylcholines.[59] While all the PCs exhibit similar temperature-dependent behavior, specific differences are observed, especially in the magnitude of the splitting for the long-chain PCs. The magnitude of the splitting at any given temperature is a result of two factors. First, a small splitting is due to a small vibrational coupling while a larger splitting indicates a larger coupling and, hence, more interchain interactions. This implies that at temperatures just below T_p, because of the small splitting observed in the CH_2 scissoring band, loose orthorhombic packing exists with a certain amount of torsional and/or librational motion about the long-chain axis. Second, perturbations to the chain packing are introduced with a reduction in chain length and presence of large terminal groups.

In summary, the packing of the lipid acyl chains in the temperature range between the pretransition (T_p) and main transition (T_m) in PCs is characterized by both X-ray and FT-IR as hexagonal, where the acyl chains may be considered to be behaving as independent rigid rotors. This type of unit cell results in little or no interchain interaction and is characterized by a CH_2 scissoring band at 1468 cm^{-1}. As the temperature is decreased below the pretransition, the acyl-chain packing is transformed into an orthorhombic-like subcell with a much greater degree of interchain vibrational coupling and hence, interchain interaction, as seen by the splitting of the CH_2 scissoring band. The fact that the magnitude of the splitting is temperature dependent indicates that the orthorhombic subcell is not a static structure, but rather contains distortions due to torsional motions about the long axis of the hydrocarbon chains.[55,60] As seen in Figure 13, as the temperature is further decreased below T_p, the factor-group splitting of the CH_2 scissoring bands for all the PCs studied increases with decreasing temperature. This indicates the gel-phase structure of the acyl chains for these lipids is becoming increasingly rigid, with little or no long axis reorientational motion.

Additional evidence for the rigid, all-trans nature of the lipid acyl chains in the gel phase comes from the spectral region 1360 to 1190 cm^{-1} containing the methylene wagging band progression (Figure 14). The band progression results from the wagging of n-coupled CH_2 oscillators in an all-trans conformation. The phase difference, ψ, allowed between the different groups (n) is given by

$$\psi = k \pi (n + 1)$$

where k is an integer from 1 to n and corresponds to the number of loops in the stationary wave that represents the normal mode.[64] The frequency pattern seen in Figure 14 is specific for an all-trans C_{16} chain.[65]

The wagging band progression is only observed in all-trans chains and, in fact, increases markedly in intensity as the temperature is decreased.[22] Its observation in the infrared spectrum of DPPC is an additional confirmation of the rigid, trans-character of the lipid acyl chains in the gel phase.

4. Interfacial Region — Carbonyl Vibrations

Of the three distinct structural domains of the phospholipid molecule — that is, the hydrophobic acyl chains, the polar head group, and the interfacial region — the least well understood in terms of its functional properties is the interfacial region. This is mainly due to the fact that the spectroscopic probes most widely used to monitor phospholipid structure are predominantly confined to probe acyl-chain behavior (e.g., ^{2}H-NMR,ESR) or head-group motions (^{31}P-NMR).[1]

The interfacial region (comprised of carbonyl esters in the most widely studied model membranes) lies at the hydrocarbon-water interface and acts as a link between the hydrophobic and hydrophilic regions of the molecule. Recent reports have suggested that due to their polar nature, the carbonyl groups at the interface might function as hydrogen-bond acceptors linking neighboring lipid molecules and that a hydrogen-bonding belt mediated by the carbonyl esters might exist along the membrane surface.[66,67] In addition, proposed models for phospholipid-cholesterol interactions describe specific hydrogen bonding between the cholesterol 3β OH and either of the phospholipid carbonyl groups.[66,68,69]

The infrared bands most characteristic of the interfacial region are the carbonyl ester C=O stretching vibrations. Levin and co-workers[70,71] have used Raman and dispersive

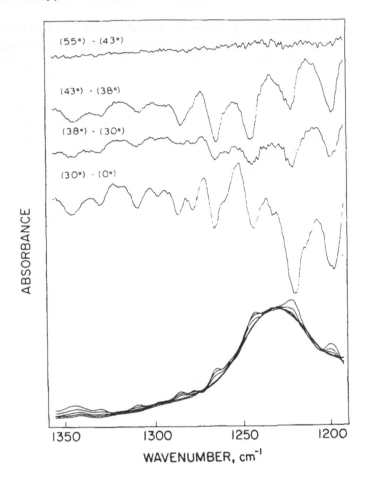

FIGURE 14.   Fourier transform infrared absorbance (bottom) and corresponding difference spectra (top, ×7) of the CH_2 wagging band progression and of the antisymmetric phosphate stretching mode of DPPC. Spectra are from a 6-μm gel and are not smoothed. (From Cameron, D. G. et al., *Biochemistry,* 20, 4496, 1981. With permission.)

IR spectroscopy to study the carbonyl groups in anhydrous phospholipids and their interaction with cholesterol. They have identified two C=O stretching modes at 1740 and 1721 cm^{-1} and have assigned these frequencies to the sn-1 and sn-2 carbonyls, respectively. The ~19 cm^{-1} difference in frequency comes about due to the fact that the sn-2 acyl chain initially packs parallel to the surface of the membrane, thus exposing it to a more hydrophilic environment, then initiates a gauche bend at the 2-position methylene that enables the sn-2 chain to pack parallel to the sn-1 chain.[29] Upon hydration, however, these features broaden and merge into a single asymmetric band contour making identification of the individual chain carbonyls difficult in the hydrated, lamellar systems commonly used as models for biological membranes.[71,72] In addition to the C=O stretching bands, the C—O stretching bands of the ester group at ~1170 and ~1070 cm^{-1} also appear in the FT-IR spectra of phospholipids. However, these bands are of limited analytical value because of vibrational coupling with C—C stretching modes as well as overlapping with the intense PO_2 stretching band at ~1080 cm^{-1}.

Until now, discussion of the thermotropic behavior of phosphocholines has concentrated on the structure of the hydrocarbon acyl chains. The IR bands characteristic of the carbonyl esters also display temperature-dependent variations which may be used to indicate structural rearrangements at the interfacial region. Due to the heavy overlap of the two C=O bands in the phospholipid spectra, Fourier self-deconvolution was used to reduce the inherent width of these bands and observe the individual carbonyls.[10,31,73] The data show that as the temperature increases, the relative peak heights of the two bands change (the sn-1 carbonyl decreases in intensity while the sn-2 carbonyl increases), whereas neither of the individual components shifts in frequency.

These changes suggest the involvement of the three-carbon glycerol backbone of the phospholipid in the phase transition process. X-ray studies have confirmed this by showing that at T_p, the angle of the tilt of the phosphocholine lipid acyl chains with respect to the normal of the bilayer decreases from 35 to 0°C.[74]

Changes in the interfacial region can also be observed during experiments designed to study gel-phase structural metastability in DPPC.[10] It has been observed that prolonged annealing of hydrated samples of DPPC at temperatures just above 0°C for an extended period of time results in a rearrangement of lipid structure.[75] Upon increasing the temperature, a new low-temperature phase transition appears which varies between 13 and 20°C for DPPC, depending on the exact thermal history of the sample.

FT-IR spectra of the C=O stretching bands of DPPC are shown in Figures 2 and 3 as a function of incubation time.[10] The data show an initial (2 to 9 hr) increase in the intensity of the band at 1741 cm^{-1} (sn-1) relative to the 1727 cm^{-1} band (sn-2). After 9 hr, however, the sn-2 carbonyl begins to increase in intensity until after 88 hr, it is ~25% more intense than the sn-1. These data indicate that incubation of DPPC at low temperature leads to dehydration of the interfacial region (and, by inference, to the head group) while the subtransition results in a head-group rehydration.

5. Head-Group Vibrations

While the vibrations discussed until now (i.e., the modes characteristic of the acyl chains and interfacial region) are common to all the phospholipids under study, the infrared bands associated with the hydrophilic head group depend on the particular class of lipids under study. These are generally the choline, ethanolamine, or serine head groups in the most widely studied lipid classes.

For the choline and ethanolamine head groups, infrared bands corresponding to several C–N modes have been identified.[76] The quaternary amine group in cholines has its CH$_3$ asymmetric stretching bands at ~3050 cm^{-1}; in addition, the methyl bending vibrations of the –N(CH$_3$)$_3^+$ group are found at 1490 and 1480 cm^{-1} (asymmetric bend). An additional analytical band is the C–N stretching band in the range 800 to 1000 cm^{-1}. For the ethanolamine head group, the –NH$_3^+$ stretching and bending vibrations are found at ~3200 and 1600 cm^{-1}, respectively. However, these are extremely weak infrared bands and are of limited diagnostic use in aqueous solution. The phosphoserine group has, in addition to those bands of the primary amine, bands characteristic of the ionized carboxylate.[77] These appear in the range 1300 to 1600 cm^{-1} for the antisymmetric and symmetric [COO]$^-$ stretching vibrations.

A series of characteristic head-group bands that are common to all the phospholipid classes are the vibrations of the phosphate ester. These occur at ~1250 cm^{-1} for the antisymmetric PO$_2^-$ stretching band and ~1080 cm^{-1} for the symmetric PO$_2^-$ stretching band. In addition, the single band P–O stretching bands are found in the 800 to 900 cm^{-1} region.[76]

Although the PO$_2^-$ stretching bands are the most intense bands in the fingerprint region of the IR spectrum, previous studies of PC and PE head groups have shown these bands to be almost invariant with temperature.[25,34]

Unlike the PC and PE head groups, the phosphatidylserine headgroup has been shown to strongly bind certain divalent cations to its headgroup area.[78] The calcium

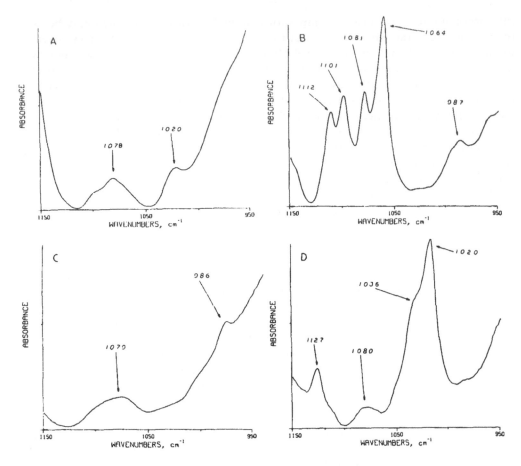

FIGURE 15.   Infrared spectra of the symmetric PO_2^- stretching region between 1150 and 950 cm^{-1} in (A) pure PS multilayers, (B) a PS-Ca^{2+} complex, (C) 0.1 M K$_2$HPO$_4$ in 0.1 M NaCl, 2 mM Tes, 2 mM His, 0.1 mM EDTA, pD 7.4, and (D) 0.1 M CaCl$_2$ plus 0.1 M K$_2$HPO$_4$ (same buffer as in [C]). (From Dluhy, R. A. et al., *Biochemistry*, 22, 6318, 1983. With permission.)

ion in particular induces a large structural reorganization which seems to be the equivalent of an isothermal phase transition from the liquid-crystalline to the crystalline state.[79] In addition, it has been proposed that the effect of Ca^{2+} is to induce a close approach of two opposing PS vesicles by displacement of water from the interbilayer space and the formation of an anhydrous, trans PS-Ca^{2+} complex spanning the interbilayer space that acts as a precursor for true membrane fusion.[80] However, in spite of the demonstrated importance of the PS head group, relatively little spectroscopic work has been done on this molecule. In particular, the site of binding of the Ca^{2+} at the PS head group has not been fully elucidated.

Dluhy et al.[26] have recently studied the binding of Ca^{2+} to PS, especially regarding the site of binding at the head group. Figure 15A shows the region 1150 to 950 cm^{-1} in the spectrum of pure PS multilayers that contains the PO_2^- symmetric stretching band superimposed upon a broad D$_2$O background. Figure 15B shows the same region in the spectrum of a PS-Ca^{2+} complex. Significant changes have occurred in the spectrum. The broad band centered at 1078 cm^{-1} in the spectrum of pure PS has split into four bands located at 1112, 1101, 1081, and 1064 cm^{-1} in that of the PS-Ca^{2+} complex.

These spectral changes may be interpreted by comparison with a simple model system. Figure 15C shows the 1150 to 950 cm^{-1} region of the spectrum of 0.1 M K$_2$HPO$_4$ in D$_2$O containing the symmetric PO_2^- stretching band. The PO_4^{3-} ion belongs to the high-symmetry point group T_d.[81] Of the fundamental vibrational modes, only ν_3 and ν_4 are IR active. When a metal ion coordinates to the free negatively charged oxygens, the symmetry is lowered by complex formation to C_{3v} or C_{2v}, and the degenerate vibrations are split.[82] This principle has been used extensively to determine the mode of coordination of acid anions such as NO_3^-, SO_4^{2-}, and CO_3^{2-} in inorganic coordination complexes.[83]

Figure 15D shows the effect on the PO_2^- stretching region of 0.1 M K_2HPO_4 of adding 0.1 M $CaCl_2$. By analogy with the spectrum of the PS-Ca^{2+} complex in Figure 15B, the lowering of the symmetry of the HPO_4^{2-} ion through coordination with Ca^{2+} has resulted in the splitting of the degenerate modes and the introduction of previously IR-inactive vibrations. As in the case of the PS-Ca^{2+} complex four bands are now evident in the spectrum. In contrast with HPO_4^{2-}, the spectrum of 0.1 M $CaCl_2$ plus 0.1 M KH_2PO_4 is indistinguishable from that of 0.1 M KH_2PO_4 alone.[26] Therefore, the fact that only the dibasic anion HPO_4^{2-} and not the monobasic anion H_2PO_4 reproduces the vibrational splitting in the presence of Ca^{2+} suggests that the Ca^{2+} ion is bound to the PS phosphate ester as a bidentate ligand.

The asymmetric PO_2^- stretching band at 1221 cm^{-1} provides further information about the environment of the phosphate ester. The band that occurs at 1221 cm^{-1} in the spectrum of pure PS shifts its frequency to 1238 cm^{-1}, decreases its FWHH, and increases intensity in the presence of Ca^{2+}. These spectral characteristics all indicate that the addition of Ca^{2+} results in the dehydration of the phosphate group.

In addition to the phosphate ester, another possible coordination site for the Ca^{2+} at the PS head group is the ionized carboxylate. However, monitoring the asymmetric [COO]$^-$ stretching band at 1622 cm^{-1} in PS before and after the addition of Ca^{2+} shows no shift in the frequency of this band, as would be expected with ion coordination. In addition, the carboxylate appears to remain hydrated.

A possible explanation for the behavior of the carboxylate in the PS=Ca^{2+} complex comes from the IR spectra of the head-group model compound, O-phospho-L-serine.[84] In this case, the infrared spectra of the asymmetric [COO]$^-$ band suggest that intermolecular hydrogen bonding may be occurring at the carboxylate. Also, the addition of equimolar Ca^{2+} does not substantially disrupt this hydrogen bonding. The data support a recent suggestion that anionic lipid head groups may be the site of a proton-conducting pathway.[85]

B. IR Spectral Studies of Natural Membranes

1. Photosensitive Membranes: Rhodopsin and Bacteriorhodopsin

Rhodopsin is the major protein (~85% in the rod outer segment) of the disk photoreceptor membrane in vertebrates and is known to be the primary light receptor. Consequently, it has been characterized structurally in some detail.[86,87] It is known that the chromophore moiety, 11-cis-retinal, is the site of the primary photochemical process and is covalently bound as a protonated Schiff base to a ε-amino lysyl side chain of the parent opsin molecule.[88-90] The absorption of a photon of light results in a cis-trans isomerization of the retinylidene chromophore and the release of retinal from its opsin matrix.

The photochemical pathway between the stable, dark-adapted 11-cis retinal and the fully bleached all-trans retinal is known to involve several intermediates existing on the pico- to microsecond time scale. Vibrational spectroscopy (both infrared and Raman) has been extensively used to characterize the photochemical cycle. Carey has recently summarized the Raman investigations.[91]

Rothschild and co-workers at Boston University have investigated the structure of bovine photoreceptor membrane using FT-IR.[92] Their studies concluded that the rhodopsin molecule contains extensive α-helical structure which is highly oriented perpendicular to the plane of the bilayer.[93,94] In addition, delipidation of the native rhodopsin leads to a change in the protein conformation. Upon reconstitution with several different lipids, however, the native structure can be completely restored.[95]

Recently, FT-IR difference spectroscopy has been used to characterize structural differences in several of the photochemical intermediates.[96] By quenching at low temperature, it is possible to stop the photochemical cycle at several of the intermediate stages. Specifically, quenching at 77 K will halt the photocycle after the first intermediate, batho-rhodopsin. The rhodopsin-bathorhodopsin difference spectrum shows that the initial photochemical intermediate involves changes in hydrogen bending and ethylenic stretching vibrations of the retinal chromophore, consistent with a cis-trans isomerization mechanism. In contrast, quenching the photocycle at a late intermediate (meta II) provides a rhodopsin-meta II difference spectrum in which the major peaks correspond to amino-acid side-chain vibrations and indicate protein conformational changes.

In addition to the rhodopsin molecule in vertebrates, another photosensitive membrane has been the object of much biophysical interest. It is the purple membrane from *Halobacterium halobium* which contains the protein bacteriorhodopsin. This bacterial pigment also contains the retinal chromophore covalently linked via a protonated Schiff base to its protein matrix. However, unlike rhodopsin, the conversion of light to chemical energy in bacteriorhodopsin serves to generate a proton gradient across the cell membrane by acting as a proton pump.

FT-IR linear dichroism studies of oriented purple membrane have been used to measure the average tilt of the α helixes in bacteriorhodopsin.[97] The Amide I, II, and A bands were shown to exhibit dichroism with vertical polarization, but not horizontal. A complete analysis of the dichroic spectra suggests that the average orientation of the α helixes in the purple membrane is less than 26° away from the normal membrane, in agreement with the model obtained from electron microscopy.[98]

The frequency of the Amide I band has been extensively used to characterize the secondary structure of proteins (see Chapter 2 of this volume). Although several physical methods indicate that the α helix is the major secondary structural component of bacteriorhodopsin,[99] IR results[100] indicate an Amide I frequency (1660 cm^{-1}) that falls outside the normal range for an α helix (1650 to 1655 cm^{-1}). Krimm and Dwivedi[101] have recently proposed, on the basis of a normal mode analysis, that this anomalous frequency is due to the helixes in bacteriorhodopsin having an αII structure rather than the usual αI. In the αII structure, the C=O group projects outward from the helix axis and is presumably stabilized by H-bonding interactions with side chains. This results in a closer contact between the amide hydrogens in the helix interior in the αII structure.

This has led to the suggestion that it is the network of NH hydrogens in the interior of the helix that leads to proton translocation across the purple membrane.[101]

2. *Acholeplasma laidlawii*

Another natural membrane which has been the subject of extensive biophysical studies is that of the microorganism *Acholeplasma laidlawii*.[102-104] *A. laidlawii* can readily be grown in strictly controlled media. Typically the medium is defatted, supplemented with a specific exogenous fatty acid, and the growth made at a specific temperature. The resultant growth is highly enriched in the exogenous acid, and the membranes exhibit a well defined gel-to-liquid-crystal phase transition at a temperature dependent on the fatty acid and the growth temperature. In recent studies the protein avidin has been added to the medium. Avidin is a potent endogenous fatty acid synthesis inhibitor and the level of exogenous fatty acid enrichment in the avidin treated membranes is generally greater than 90%.

Most biophysical studies have been concerned with the nature of the phase transi-

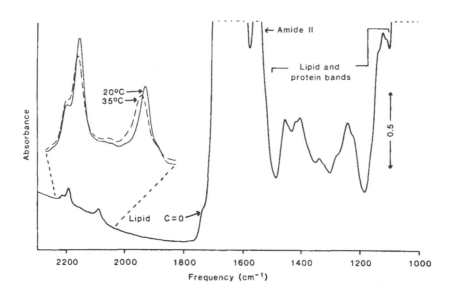

FIGURE 16. Infrared spectrum of *A. laidlawii B* grown at 30°C and recorded at 30°C. The C–D stretching region at 20 and 35°C shows the changes resulting from transition of the membrane lipids from the gel phase (20°C) to the liquid-crystal phase (35°C). (From Cameron, D. G. et al., *Science*, 219, 180, 1983. With permission.)

tion.[105-108] FT-IR shows great promise in such studies. The microorganisms can readily be grown on deuterated fatty acids. When these are incorporated into the membrane lipids, the spectra of the resultant intact and isolated membranes show well-defined C–D stretching bands in the 2200 to 2100 cm^{-1} region of the spectrum (see Section III.A.2). Figure 16 shows an example of such a spectrum. *A. laidlawii B* were grown on perdeuteromyristic acid in the presence of avidin at 30°C. The spectrum is that of the live organism and was recorded in 10 sec. The insert shows in detail the C–D stretching region of the spectrum at 20 and 35°C. The 20°C spectrum is that of the gel phase while the 35°C spectrum is that of the liquid-crystal phase. The changes in the C–D stretching bands are quite apparent and similar to those observed in studies of synthetic membranes.[9,47-49]

Earlier studies of Casal et al.[109-111] were concerned with comparisons of the phase transitions of intact membranes, obtained by lysis and washing, with those of the lipids extracted from the membranes. They addressed the question of whether the proteins had any substantial effect on the phase transition of the membrane lipids. The temperatures and characteristics of the transitions were virtually identical. However, the frequency of the C–D stretching bands was higher in the liquid-crystal phase of the intact membranes than in that of the isolated membranes, suggesting that the proteins increase the fluidity in the liquid-crystal phase.

More recent studies have been concerned with the nature of the transition in the membranes of living organisms.[112] An implicit assumption in studies of isolated membranes is that the results are relevant to the live cell. However, some 2H-NMR studies had shown that the isolated membranes of *A. laidlawii* grown on saturated fatty acid were virtually completely in the gel phase at the growth temperatures.[113,114] This was incompatible with the observation that enzyme activity was highly dependent on the degree of fluidity of the lipid membrane. In the first study *A. laidlawii B* were grown at 30°C on perdeuteromyristic acid with avidin present. The resultant membranes exhibited a phase transition in the range 22 to 35°C (Figure 17) and, as measured by the two-component model (Section III.A.2) were 20% liquid-crystalline phase at 30°C. However, the live cells exhibited a lower temperature phase transition (Figure 17) and

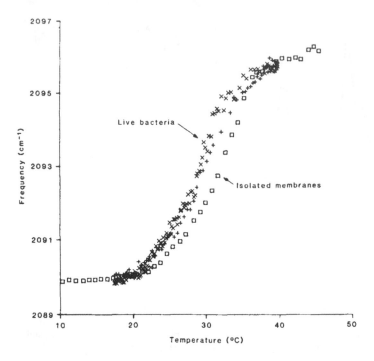

FIGURE 17. Temperature dependence of the frequency of the CD_2 symmetric stretching band of the lipids of *A. laidlawii B* grown at 30°C on perdeuteromyristic acid in the presence of avidin. Shown are frequencies from spectra of live cells with the temperature ascending from 20 to 39°C (+) and descending from 39 to 16°C (X) and frequencies from spectra of isolated membranes with the temperature ascending from 5 to 45°C (□). Frequencies were determined by computing the center of gravity of the topmost 2 cm^{-1} segment of the band. (From Cameron, D. G. et al., *Science*, 219, 180, 1983. With permission.)

were 50% liquid-crystalline phase at the growth temperature. The studies have since been extended to growths in the range 25 to 37°C with myristic, pentadecanoic, and palmitic exogenous fatty acids.[115] In all cases the live cells are more fluid at the growth temperature than are the isolated membranes. The results thus confirm those of the NMR studies. However, they also indicate that studies of isolated membranes are not always directly relevant to the properties of live cells. It would appear that either the membrane isolation alters the composition of the membranes sufficiently to shift the phase transition, or that the live organism is able to actively increase the degree of fluidity of the membrane.

Finally, the spectrum in Figure 16 has a large number of bands in the region 1800 to 1000 cm^{-1} which are from the proteins, lipids, and other molecules comprising the organism. These have not been studied to date, and surely have the potential to lead to further insights into the nature of the living organism.

IV. CONCLUSION

The data presented in this chapter and other chapters in this volume serve to indicate that the conditions that have prevented the application of IR spectroscopy to biochemistry in general, and membrane biophysics in particular, have largely been overcome. The development of Fourier transform instrumentation with its enhanced sensitivity, extremely high precision, and dedicated computers has made water subtraction a routine operation. Perhaps more importantly, the development of algorithms that reduce the inherent width of the broad liquid-phase IR bandshapes has enabled investigators to uncover previously hidden structure in the measured IR bands. As the instrumentation evolves and more investigators become aware of the potential of biochemical FT-IR spectroscopy, the next several years should see its increased application to a wide variety of biochemical problems.

ABBREVIATIONS

ATR — Attenuated total reflectance

DMPC — 1,2-Dimyristol-sn-glycero-3-phosphocholine

DMPC-d_{54} — Acyl chain perdeuterated DMPC

DOPC — 1,2-Dioleoyl-sn-glycero-3-phosphocholine

DPPC — 1,2-Dipalmitoyl-sn-glycero-3-phosphocholine

DPPC-d_{62} — Acyl chain perdeuterated DPPC

DPPE — 1,2-Dipalmitoyl-sn-glycro-3-phosphoethanolamine

ESR — Electron spin resonance

FT-IR — Fourier transform infrared spectroscopy

FWHH — Full width at half height

ILSF — Instrument line shape function

IR — Infrared spectroscopy

NMR — Nuclear magnetic resonance

PS — Bovine brain phosphatidylserine

SNR — Signal-to-noise ratio

T_M — Main gel to liquid-crystalline phase transition

T_P — Temperature of the gel phase pretransition

REFERENCES

1. Grell, E., Ed., *Membrane Spectroscopy*, Springer-Verlag, Basel, 1981.
2. Wilson, E. B., Decius, J. C. and Cross, P. C., *Molecular Vibrations*, McGraw-Hill, New York, 1955.
3. Lipp, E. and Nafie, L., *Appl. Spectrosc.*, 38, 20, 1984.
4. Lipp, E., Zimba, C. G., and Nafie, L., *Chem. Phys. Lett.*, 90, 1, 1982.

5. Nafie, L. and Vidrine, W., in *Fourier Transform Infrared Spectroscopy*, Vol. 3, Ferraro, J. R. and Basile, L. J., Eds., Academic Press, New York, 1982, 83.
6. Savitzky, A. and Golay, M., *Anal. Chem.*, 36, 1627, 1964.
7. Cameron, D. G., Kauppinen, J. K., Moffatt, D., and Mantsch, H., *Appl. Spectrosc.*, 36, 245, 1983.
8. Umemura, J., Mantsch, H., and Cameron, D. G., *J. Colloid Interface Sci.*, 83, 558, 1981.
9. Dluhy, R. A., Mendelsohn, R., Casal, H. L., and Mantsch, H., *Biochemistry*, 22, 1170, 1983.
10. Cameron, D. and Mantsch, H., *Biophys. J.*, 38, 175, 1982.
11. Kauppinen, J. K., Moffatt, D. J., Mantsch, H. H., and Cameron, D. G., *Appl. Spectrosc.*, 35, 271, 1981.
12. Kauppinen, J. K., Moffatt, D. J., Cameron, D. G., and Mantsch, H. H., *Appl. Optics*, 30, 1866, 1981.
13. Kauppinen, J. K., Moffatt, D. J., Mantsch, H. H., and Cameron, D. G., *Anal. Chem.*, 53, 1454, 1981.
14. Kauppinen, J. K., Moffatt, D. J., Mantsch, H. H., and Cameron, D. G., *Appl. Optics*, 21, 1866, 1982.
15. Cameron, D. G. and Moffatt, D. J., *J. Test. Eval.*, 12, 78, 1984.
16. Pitha, J. and Jones, R. N., *Can. J. Chem.*, 44, 3031, 1966.
17. Fraser, R. D. B. and Suzuki, E., *Anal. Chem.*, 38, 1770, 1966.
18. Fraser, R. D. B. and Suzuki, E., *Anal. Chem.*, 41, 37, 1969.
19. Aloia, R. C., Ed., *Membrane Fluidity in Biology*, Vol. 2, Academic Press, New York, 1983.
20. Cameron, D. B., Casal, H. L., and Mantsch, H. H., *J. Biochem. Biophys. Metho.*, 1, 21, 1979.
21. Snyder, R. G., Strauss, H. L., and Elliger, C. A., *J. Phys. Chem.*, 86, 5145, 1982.
22. Cameron, D. G., Casal, H. L., and Mantsch, H. H., *Biochemistry*, 19, 3665, 1980.
23. Cameron, D. G. and Mantsch, H. H., *Biochem. Biophys. Res. Commun.*, 83, 886, 1978.
24. Umemura, J., Cameron, D. G., and Mantsch, H. H., *Biochim. Biophys. Acta*, 602, 32, 1980.
25. Mendelsohn, R., Dluhy, R., Taraschi, T., Cameron, D. G., and Mantsch, H. H., *Biochemistry*, 20, 6699, 1981.
26. Dluhy, R. A., Cameron, D. G., Mantsch, H. H., and Mendelsohn, R., *Biochemistry*, 22, 6318, 1983.
27. Casal, H. L. and Mantsch, H. H., *Biochim. Biophys. Acta*, 779, 555, 1984.
28. Chapman, D., Urbina, J., and Keough, K. M., *J. Biol. Chem.*, 249, 2512, 1974.
29. Pearson, R. and Pascher, I., *Nature (London)*, 281, 499, 1979.
30. Hauser, H., Pascher, I., Pearson, R., and Sundell, S., *Biochim. Biophys. Acta*, 650, 21, 1981.
31. Casal, H. L. and Mantsch, H. H., *Biochim. Biophys. Acta*, 735, 387, 1983.
32. Luzatti, V., Gulik-Krzywicki, T., and Tardieu, A., *Nature (London)*, 218, 1031, 1968.
33. Cullis, R. P. and de Kruiff, B., *Biochim. Biophys. Acta*, 559, 399, 1979.
34. Mantsch, H. H., Martin, A., and Cameron, D. G., *Biochemistry*, 20, 3138, 1981.
35. Parsegian, A., Ed., *Biophysical Discussions; Protein-Lipid Interactions in Membranes*, Rockefeller University Press, New York, 1981.
36. Mendelsohn, R., Dluhy, R. A., Crawford, T., and Mantsch. H. H., *Biochemistry*, 23, 1498, 1984.
37. MacLennan, D. H. and Holland, P. C., in *The Enzymes of Biological Membranes*, Vol. 3, Martonosi, A., Ed., Plenum Press, New York, 1976, 221.
38. Hildago, C., Ikemoto, N., and Gergely, J., *J. Biol. Chem.*, 251, 4224, 1976.
39. Mendelsohn, R., Anderle, G., Jaworsky, M., Mantsch, H., and Dluhy, R. A., *Biochim. Biophys Acta*, 775, 215, 1984.
40. Klausner, R. D., Kleinfeld, A. M., Hoover, R. L., and Karnovsky, M. J., *J. Biol. Chem.*, 255, 1286, 1980.
41. Lee, A. G., *Biochim. Biophys. Acta*, 472, 285, 1977.
42. Smith, I. C. P. and Mantsch, H. H., *Trends Biochem. Sci.*, 4, 152, 1979.
43. Mendelsohn, R. and Koch, C., *Biochim. Biophys. Acta*, , 1980.
44. Sunder, S., Cameron, D. G., Mantsch, H. H., and Bernstein, H. J., *Can. J. Chem.*, 56, 2121, 1978.
45. Sunder, S., Mendelsohn, R., and Bernstein, H., *Chem. Phys. Lipids*, 17, 456, 1976.
46. Bunow, M. and Levin, I. W., *Biochim. Biophys. Acta*, 489, 191, 1977.
47. Dluhy, R. A., Moffatt, D. J., Cameron, D. G., Mendelsohn, R., and Mantsch, H. H., *Can. J. Chem.*, in press, 1985.
48. Mendelsohn, R. and Maisano, J., *Biochim. Biophys. Acta*, 506, 191, 1979.
49. Mendelsohn, R., Brauner, J., Faines, L., Mantsch, H. H., and Dluhy, R. A., *Biochim. Biophys. Acta*, 774, 237, 1984.
50. Singer, M., *Chem. Phys. Lipids*, 31, 145, 1982.
51. Tse, M. Y. and Singer, M., *Can. J. Biochem.*, 62, 72, 1984.
52. Mabrey, S. and Sturtevant, J., in *Methods in Membrane Biology*, Korn, E., Ed., Plenum Press, New York, 1978.
53. Reisman, A., *Phase Equilibria*, Academic Press, New York, 1970.
54. Seelig, J., *Q. Rev. Biophys.*, 10, 353, 1977.

55. Davis, J. H., *Biophys. J.*, 27, 339, 1979.
56. Cameron, D. G., Casal, H. L., Mantsch, H. H., Boulanger, Y., and Smith, I. C. P., *Biophys. J.*, 35, 1, 1981.
57. Sunder, S., Cameron, D. G., Casal, H. L., Boulanger, Y., and Mantsch, H. H., *Chem. Phys. Lipids*, 28, 137, 1981.
58. Cameron, D. G., Casal, H. L., Gudgin, E. F., and Mantsch, H. H., *Biochim. Biophys. Acta*, 596, 463, 1980.
59. Cameron, D. G., Gudgin, E. F., and Mantsch, H. H., *Biochemistry*, 20, 4496, 1981.
60. Janiak, M. J., Small, D. M., and Shipley, G. G., *J. Biol. Chem.*, 254, 6068, 1979.
61. Snyder, R. G., *J. Mol. Spectrosc.*, 1, 116, 1961.
62. Snyder, R. G., *J. Chem. Phys.*, 71, 3229, 1979.
63. Decius, J. C. and Hexter, M., *Molecular Vibrations in Crystals*, McGraw-Hill, New York, 1977.
64. Snyder, R. G. and Schachtschneider, J. H., *Spectrochim. Acta*, 19, 85, 1963.
65. Susi, H. and Panzer, S., *Spectrochim. Acta*, 18, 499, 1962.
66. Brockerhoff, H., *Lipids*, 9, 645, 1974.
67. Ramsammy, L. S., Chauhan, V. P. S., Box, L. L., and Brockerhoff, H., *Biochem. Biophys. Res. Commun.*, 118, 743, 1984.
68. Huang, C., *Lipids*, 12, 348, 1977.
69. Huang, C., *Nature (London)*, 259, 242, 1976.
70. Bush, S. F., Adams, R. G., and Levin, I. W., *Biochemistry*, 19, 4429, 1980.
71. Bush, S. F., Levin, H., and Levin, I. W., *Chem. Phys. Lipids*, 27, 101, 1980.
72. Mushayakarara, E., Albon, N., and Levin, I. W., *Biochim. Biophys. Acta*, 686, 153, 1982.
73. Mantsch, H. H., Cameron, D. G., Tremblay, P. A., and Kates, M., *Biochim. Biophys. Acta*, 689, 63, 1982.
74. Rand, R. P. D., Chapman, D., and Larsson, K., *Biophys. J.*, 15, 1117, 1975.
75. Chen, S. C., Sturtevant, J. M., and Gaffney, B. J., *Proc. Natl. Acad. Sci., U.S.A.*, 77, 5060, 1980.
76. Fringeli, U. P. and Gunthard, H., in *Membrane Spectroscopy*, Grell, E., Ed., Springer-Verlag, Basel, 1981, 270.
77. Bellamy, L. J., *The Infra-red Spectra of Complex Molecules*, Chapman and Hall, London, 1975.
78. Poste, G. and Allison, A. C., *Biochim. Biophys. Acta*, 300, 421, 1973.
79. Papahadjopoulos, D., Vail, W. J., Newton, C., Nir, S., Jacobson, K., Poste, G., and Lazo, R., *Biochim. Biophys. Acta*, 465, 579, 1977.
80. Portis, A., Newton, C., Pangborn, W., and Papahadjopoulos, D., *Biochemistry*, 18, 780, 1979.
81. Cotton, F. A. and Wilkinson, G., *Advanced Inorganic Chemistry*, Interscience, New York, 1972.
82. Lincoln, S. F. and Stranks, D. R., *Aust. J. Chem.*, 21, 37, 1968.
83. Nakamoto, K., *Infrared and Raman Spectra of Inorganic and Coordination Compounds*, Interscience, New York, 1978.
84. Dluhy, R. A., Cameron, D. G., and Mantsch, H. H., *Biochim. Biophys. Acta*, 792, 182, 1984.
85. Haines, T., *Proc. Natl. Acad. Sci. U.S.A.*, 80, 160, 1983.
86. Callender, R. and Honig, B., *Ann. Rev. Biophys. Bioeng.*, 6, 33, 1977.
87. Mathies, R., in *Chemical and Biochemical Applications of Lasers*, Vol. 4, Moore, C. B., Ed., Academic Press, New York, 1977, 55.
88. Lewis, A., *Bull. Am. Phys. Soc. II*, 19, 375, 1974.
89. Callender, R., Daukas, A., Crouch, R., and Nakanishi, K., *Biochemistry*, 15, 1621, 1976.
90. Mathies, R., Oseroff, A. R., and Stryer, L., *Proc. Natl. Acad. Sci., U.S.A.*, 73, 1, 1976.
91. Carey, P., *Biochemical Applications of Raman and Resonance Raman Spectroscopies*, Academic Press, New York, 1982.
92. Rothschild, K. J., Sanches, R., and Clark, N. A., *Meth. Enzymol.*, 88, 696, 1982.
93. Rothschild, K. J., Clark, N. A., Rosen, K. M., Sanches, R., and Hsiao, T. L., *Biochem. Biophys. Res. Commun.*, 92, 1266, 1980.
94. Rothschild, K. J., Sanches, R., Hsiao, T. L., and Clark, N. A., *Biophys. J.*, 31, 53, 1980.
95. Rothschild, K. J., Degrip, W. J., and Sanches, R., *Biochem. Biophys. Acta*, 596, 338, 1980.
96. Rothschild, K. J., Cantore, W. A., and Marrero, H., *Science*, 219, 1333, 1983.
97. Rothschild, K. J. and Clark, N. A., *Biophys. J.*, 25, 473, 1979.
98. Henderson, R. and Univin, P. N. T., *Nature (London)*, 257, 28, 1975.
99. Blaurock, A., *J. Mol. Biol.*, 93, 139, 1975.
100. Rothschild, K. J. and Clark, N. A., *Science*, 204, 311, 1979.
101. Krima, S. and Dwived, A. M., *Science*, 216, 407, 1982.
102. McElhaney, R., *Biochim. Biophys. Acta*, 779, 1, 1984.
103. Wieslander, A., Rilfors, L., Johansson, L. B. A., and Lindbloom, G., *Biochemistry*, 20, 730, 1981.
104. Stockton, G. W., Johnson, K. G., Butler, K. W., Tulloch, A. P., Boulanger, Y., Smith, I. C. P., Davis, J. H., and Bloom, M., (1979)

105. Steim, J. M., Tourtellotte, M. E., Reinert, J. C., McElhaney, R. N., and Rader, R. L., *Proc. Natl. Acad. Sci., U.S.A.*, 63, 104, 1969.
106. Engleman, D., *J. Mol. Biol.*, 47, 115, 1970.
107. Smith, I. C. P., Butler, K. W., Tulloch, A. P., Davis, J. H., and Bloom, M., *FEBS Lett.*, 100, 57, 1979.
108. Butler, K. W., Johnson, K. G., and Smith, I. C. P., *Arch. Biochem. Biophys.*, 191, 289, 1978.
109. Casal, H. L., Smith, I. C. P., Cameron, D. G., and Mantsch, H. H., *Biochim. Biophys. Acta*, 550, 145, 1979.
110. Casal, H. L., Cameron, D. G., Smith, I. C. P., and Mantsch, H. H., *Biochemistry*, 19, 444, 1980.
111. Casal, H. L., Cameron, D. G., Jarrell, H. C., Smith, I. C. P., and Mantsch, H. H., *Chem. Phys. Lipids*, 30, 17, 1982.
112. Cameron, D. G., Martin, A., and Mantsch, H. H., *Science*, 219, 180, 1983.
113. Jarrell, H. C., Butler, K. W., Byrd, R. A., Deslauriers, R., Ekiel, I., and Smith, I. C. P., *Biochim. Biophys. Acta*, 688, 622, 1982.
114. Kang, S. Y., Kinsey, R. A., Rajan, S., Gutowsky, H. S., Gabridge, M. G., and Oldfield, E., *J. Biol. Chem.*, 256, 1155, 1981.
115. Cameron, D. G., Martin, A., and Mantsch, H. H., manuscript in preparation, 1985.

Chapter 4

FOURIER TRANSFORM MASS SPECTROMETRY

Alan G. Marshall

TABLE OF CONTENTS

I. PRINCIPLES OF FOURIER TRANSFORM MASS SPECTROMETRY (FT-MS)

Fourier transform methods offer three main advantages for spectroscopy.[1,2] First, multiplex detection followed by Fourier "decoding" leads to a multiplex or Fellgett advantage of a factor of up to $\sqrt{N}$ in signal-to-noise ratio (or $1/N$ in time) compared to an instrument which scans one data point at a time, where N is the number of data points in the frequency spectrum. When (as in FT-NMR, FT-IR, and FT-MS) a spectral data set consists of about 10,000 points, these advantages translate to improvements of a factor of about 10,000 in speed or 100 in signal-to-noise ratio (SNR) over conventional scanning instruments. It is worth noting that an additional advantage unique to FT-MS is that spectral resolution also increases by a factor of about 5000 over the corresponding continuous-wave scanning instrument.

Second, Fourier techniques provide a variety of simple means for manipulating digitized data: smoothing or filtering to enhance SNR; resolution enhancement (via apodization to narrow the spectral peak, or zero-filling to give more points per line width); changing the spectral shape to reduce peak overlap (e.g., from Lorentzian to Gaussian shape); generation of integrals or derivatives; and clipping to reduce spectral storage requirements. Third, for any linear system (i.e., output is proportional to input), Fourier methods can be used to remove any known irregularities in the excitation waveform, so that the corrected ("deconvoluted") response reflects only the properties of the sample, and not the imperfections of the measuring device.

In this chapter we begin by reviewing the origin, form, excitation, and detection of ion cyclotron resonance mass spectra. The effect and interplay of various experimental parameters is discussed in Sections I.A. through I.D. Finally, current examples and prospects for biomedical applications of FT-MS are presented in Sections II.A. through II.K.

A. Excitation/Detection

Left by itself, a moving ion in a static magnetic field is bent into a circular path. The angular velocity of the ion motion is given by

$$\omega = \frac{qB}{m} \tag{1}$$

in which q/m is the charge-to-mass ratio (C/kg) of the ion, and B is the applied magnetic field strength (tesla, T). It is therefore clear that if the magnetic field strength were known, measurement of the ion "cyclotron" frequency of Equation 1 would suffice to determine q/m. In other words, a static magnetic field effectively "converts" ionic mass into frequency. For a typically available magnet (about 3 T), the cyclotron frequencies for singly charged ions range from a few kilohertz up to a few megahertz (see Figure 1); these radiofrequencies can be measured with great precision.

However, when ions are formed at random time intervals (as in all present applications), ions of a given initial velocity will be distributed randomly about their circle, and their "incoherent" motion does not yet produce an external signal. In order to detect the ions, it is necessary to move them "off center". A simple method is to form the ions in the center of a rectangular box of metal plates (e.g., with a short [<100 msec] beam of 50-eV electrons), and then accelerate the ions with a rotating electric field. After a few cycles, the ions form a packet which spirals outward with time, and the motion is described as "ion cyclotron resonance" (ICR).

In continuous-wave ICR spectrometers,[3] spiral motion at a given q/m ratio was detected from the power absorbed at that frequency during the excitation (Figure 2a).

$$\nu = \frac{qB}{2\pi m} \ , \text{in kHz}$$

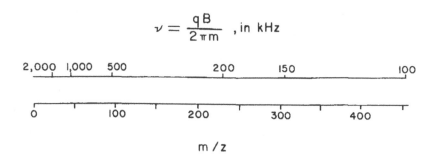

m / z

FIGURE 1. Nomogram relating mass and frequency scales in ion cyclotron resonance mass spectrometry, at an applied magnetic field strength of 3.0 T.

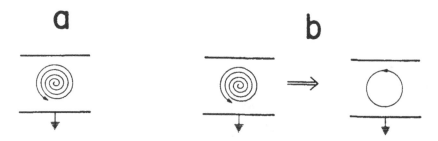

FIGURE 2. Theoretical ion cyclotron trajectory during excitation and detection. Magnetic field direction is perpendicular to the page. (a) Continuous-wave ICR excitation — detection is based on resonant power absorption during the excitation period; (b) Fourier-transform ICR — the ion is first excited to a "parking" orbit, and the circular motion then detected from the oscillating charge induced in a pair of opposing plates enclosing the sample.

A spectrum was obtained by slowly (about 20 min) scanning the magnetic field strength, much the same as in field-sweep nuclear magnetic resonance (NMR) spectrometers. In Fourier transform ICR spectrometers,[4,5] ions are first excited to a predetermined radius, and then "parked" in that orbit by turning off the rf electric field excitation (Figure 2b). The orbiting ion packet then induces an oscillating charge in the surrounding metal plates; that charge is converted to a voltage, amplified, digitized, and stored in a computer, all in about 0.1 sec.

When ions of many q/m ratios are present, it can be shown that excitation with rf power the amplitude of which is constant with frequency will excite all the ions to orbits of the same radius.[6] When the rf excitation is turned off, all the ICR frequencies will be "heard" by the detector plates, just as all the frequencies from an orchestra are heard simultaneously. A Fourier transform of the digitized time-domain signal from the detector plates then yields a frequency-domain spectrum, which can readily be rescaled as a mass spectrum. A fuller description of the general principles and advantages of Fourier transform spectrometry may be found in References 1 and 2.

B. Spectral Line Shapes

A complex Fourier transform of a time-domain ICR signal yields two types of spectra: absorption and dispersion (Figure 3).[7] The shape of the spectral lines depends on two parameters: the length of the observation period and the time constant for disappearance of coherent ion motion (due to ion-molecule collisions or reactions).[8] In general, the longer the time-domain signal persists, the narrower is the frequency-domain peak, and the higher the frequency (or mass) resolution. Thus, an important prior

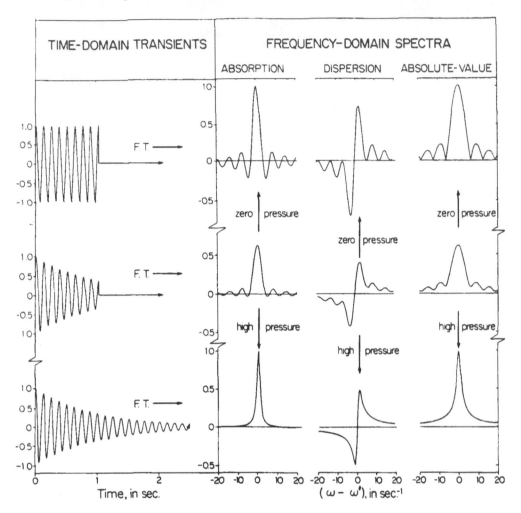

FIGURE 3. Cosine transform (absorption), sine transform (dispersion), and magnitude (absolute-value) frequency-domain ICR spectra obtained by analytical Fourier transformation of the continuous time-domain signals shown at the left of each set of spectra. The top row corresponds to the zero-pressure limit (i.e., no collisions during the observation period), and the bottom row to the high-pressure limit (i.e., ICR signal is collisionally damped to zero during the observation period). The middle row represents an intermediate pressure for which the observation period lasts for one exponential damping time constant. (From Marshall, A. G., Comisarow, M. B., and Parisod, G., *J. Chem. Phys.*, 71, 4434, 1979. With permission.)

development was McIver's trapped-ion cell, which increased the ion-observation period from a few milliseconds to several hours.[9]

In practice, recovery of the absorption signal $A(\omega)$ or the dispersion signal $D(\omega)$ requires phase corrections which vary nonlinearly with frequency.[10,11] Therefore, FT-MS spectra are normally reported as "magnitude-mode", $M(\omega)$,

$$M(\omega) = [[A(\omega)]^2 + [D(\omega)]^2]^{1/2} \qquad (2)$$

which is independent of phase. As seen in Figure 3, the magnitude-mode signal is broader than the absorption signal by a factor of between $\sqrt{3}$ and 2.

1. Lorentz Limit

In the extreme limit that the ions are observed for an infinite period, so that the ICR signal has decreased (exponentially) to zero, the frequency-domain line shape will be Lorentzian, as shown in the lowermost plots in Figure 3.[8]

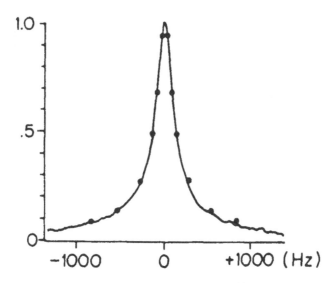

FIGURE 4. Experimental (dots) and theoretical (solid curve) magnitude-mode FT/ICR mass spectrum for CH_5^+ ions produced by electron ionization of methane at 2×10^{-5} torr. The solid curve corresponds to an observation period of 40.96 msec, and a time-domain damping constant of 2.1 msec. (From Marshall, A. G., Comisarow, M. B., and Parisod, G., *J. Chem. Phys.*, 71, 4434, 1979. With permission.)

2. Sinc Limit

In the opposite extreme limit that the ICR signal has not decayed at all during the observation period, the frequency-domain line shape is the same "sinc" function that describes diffraction from a slit, as shown in the topmost plots in Figure 3.[12]

3. General Case

In general, observations will fall between the two extremes just discussed, and the line shape for a typical intermediate case is shown in the middle row of plots in Figure 3. Although the frequency-sweep excitation[3] on which present commercial FT-MS instruments are based produces a somewhat more complicated line shape,[11] the qualitative features of Figure 3 are still preserved. The close agreement between experimental and theoretical line shapes in Figure 4 supports the validity of the theory.[8]

C. Mass Resolution

Mass resolution in FT-MS is determined by the "analog" line width of the frequency-domain signal (which is in turn determined by the sample pressure), and the "digital" resolution (i.e., the number of discrete data points per analog line width). It can be shown that analog frequency resolution and mass resolution are identical,[12] so that the frequency-domain descriptions of Figure 3 suffice to predict mass resolution in FT-MS.

1. Analog Mass Resolution
a. Dependence on Pressure and Length of Observation Period

For an ICR time-domain signal which decays exponentially with a time constant τ, the minimum possible FT-MS absorption-mode line width is $(1/\pi\tau)$ Hz, corresponding to an infinite observation period. As the observation period is shortened, the line width increases (and the mass resolution decreases),[8,13] as shown in Figure 5. Figure 6 shows experimental and theoretical mass resolution as a function of observation period, for

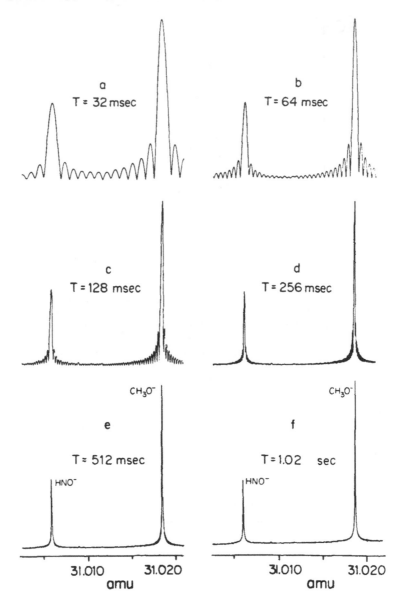

FIGURE 5.   Experimental FT-ICR magnitude-mode mass spectra of HNO⁻ and
CH₃O⁻ as a function of data-acquisition period. A time-domain transient ICR
signal from HNO⁻ and CH₃O⁻ was digitized and acquired for periods ranging
from 32 msec (a) to 1.02 sec (f). The line widths of the CH₃O⁻ spectra shown
here furnish the data for the next figure. (From Marshall, A. G., Comisarow,
M. B., and Parisod, G., *J. Chem. Phys.*, 71, 4434, 1979. With permission.)

fixed pressure. Since τ varies inversely with pressure, Figure 6 is easily rescaled to
correspond to any other pressure.

b. Dependence on Mass

For equal time-domain observation period and damping constant, ions of various
q/m ratios will have equal frequency-domain line widths in the FT-MS experiment.[8]
However, since ICR frequency varies inversely with ionic mass (see Equation 1), FT-
ICR line width in mass units varies as m^2, so that mass resolution varies inversely with
mass, as seen from the experiment/theory comparison of Figure 7. Even so, mass

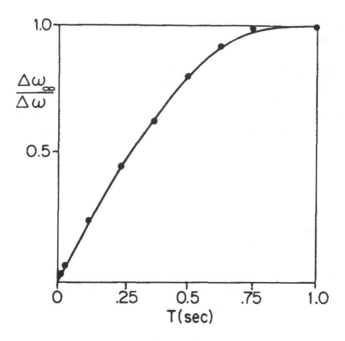

FIGURE 6. Dependence of the experimental FT-ICR collision-broadened line width upon data-acquisition period. Each solid circle represents the ratio of the line width for maximal acquisition period to the line width at the stated (shorter) acquisition period on the abscissa. The solid curve is the theoretical maximum relative mass resolution computed from the line width at maximal acquisition period. (From Marshall, A. G., Comisarow, M. B., and Parisod, G., *J. Chem. Phys.*, 71, 4434, 1979. With permission.)

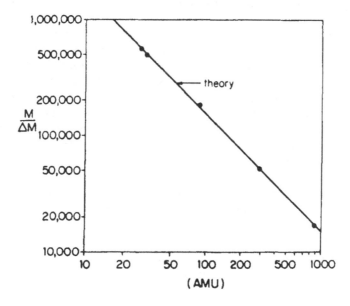

FIGURE 7. Mass resolution as a function of ionic mass. Straight line, theory; solid circles, experimental data for N_2^+, $C_2H_4^+$, NOH^-, CH_3O^-, $^{84}Kr^+$, $C_6H_{12}^+$, dibromoadamantane (positive parent ions), and $C_{18}F_{32}N_3^+$. (From Comisarow, M. B., in *Fourier, Hadamard, and Hilbert Transforms in Chemistry*, Marshall, A. G., Ed., Plenum Press, New York, 1982, 133. With permission.)

resolution exceeding 100,000:1 is readily obtained at 1,000 atomic mass unit (amu) for singly charged ions.

2. Digital Mass Resolution: Dependence on Size of Data Set

It is obvious that the precision with which a mass spectral peak can be located is directly related to the number of data points per unit mass. For an observation period of T sec, the frequency-domain points will be spaced at intervals of 1/T Hz. The maximum possible T is in turn determined by the maximum available number of time-domain data points, N:

$$F \cdot T = N \tag{3}$$

The sampling rate, F, must be at least twice the maximum ICR frequency (if the time-domain signal is digitized directly) or twice the heterodyne bandwidth (if the time-domain signal is first heterodyned against a reference oscillator before digitization).[1]

In practice, therefore, one first chooses the desired mass range. For direct detection, F is simply twice the ICR frequency of the smallest ionic mass in the range; for heterodyne detection, F is twice the frequency bandwidth corresponding to the mass range of interest. N then determines T. Zero-filling provides a way to increase the *digital* resolution (i.e., by increasing N after data collection);[14] however, zero-filling does not reduce the *analog* line width, which is determined by T and τ.

D. Signal-To-Noise Ratio

1. Signal Averaging

When, as in ICR, the noise is independent of the signal strength, the SNR obtained by adding n individual time- or frequency-domain data sets will be higher by a factor of $n^{1/2}$ than the SNR for a single scan.[1,15] Since several thousand scans can be accumulated in a few minutes, signal averaging can increase the SNR in FT-MS by 100-fold or more compared to a single-scan spectrum.

2. Trade-Offs Between Signal-To-Noise Ratio and Mass Resolution

In a magnetic sector mass spectrometer, mass resolution is increased by narrowing the analyzer slit width, thereby decreasing signal-to-noise ratio. In FT-MS, on the other hand, several different trade-offs between SNR and resolution can be obtained from a single stored data set. SNR may be enhanced at the expense of resolution by preferentially weighting ("apodizing") the first portion of the time-domain data, whereas resolution may be enhanced at the expense of SNR by preferentially weighting the last portion of the time-domain data.[15] It is as if the FT-MS instrument operates with a software-adjustable "slit width".

In choosing the conditions for spectral acquisition, Equation 3 shows that for highest mass resolution, maximum N gives maximum T. If signal-to-noise is more important than resolution, T per scan may be made shorter by choosing smaller N. The product of SNR and resolution is optimal for $T \cong 3\tau$, at which mass resolution has reached about 92% of its limiting value, while the SNR retains more than 30% of its theoretically maximum value.[16]

Finally, since SNR is proportional to the number of ions, and since FT-MS line width varies inversely with pressure, it is possible to reduce the pressure (thereby increasing the mass resolution) without loss of signal-to-noise, provided that the same number of ions is generated (e.g., by increasing the filament current as the pressure is lowered).[17] Thus, it may not even be necessary to reduce SNR in order to improve resolution.

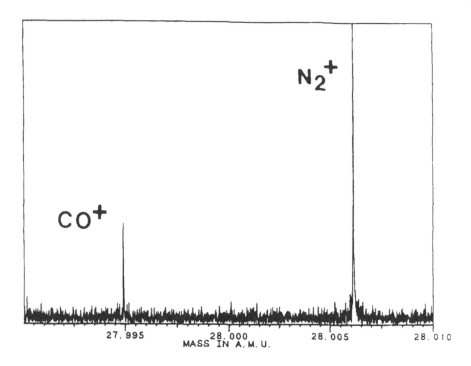

FIGURE 8.   Ultrahigh resolution (1,147,000:1) Fourier transform mass spectrum from background nitrogen and carbon monoxide, obtained with a Nicolet FTMS-1000 spectrometer. The two peaks differ in mass by 1 part in 2493.

II. PERFORMANCE FEATURES AND APPLICATIONS

A. Ultrahigh Mass Resolution and Speed

Figure 8 shows a Fourier transform positive-ion mass spectrum of residual nitrogen and carbon monoxide at a spectrometer base pressure of about 2×10^{-9} torr. Although mass resolution varies inversely with mass (Figure 7), the resolution observed in Figure 8 corresponds to a theoretical mass resolution of >30,000 at m/z = 1000. In practice, line widths of about 0.1 amu at m/z = 1000 or higher are routinely obtained with solids probe samples.[18] It is worth noting that even an ultrahigh resolution spectrum can be obtained in a short time: 1.4 sec in the example of Figure 8.

B. Positive/Negative Ion Detection

Ions in the cyclotron are constrained by the magnetic field against escape along any path perpendicular to the magnetic field. In order to prevent ions from escaping along a path parallel to the magnetic field, a small "trapping" potential of about 1 V/cm is applied to plates which form the two ends of the ICR excitation/detection "cell". In order to change from positive ion detection to negative ion detection, one simply changes from a positive to a negative trapping potential. Figures 5 and 11 show negative ion spectra; the remaining experimental spectra show positive ions.

C. Mass Calibration: Exact Mass Determination

In practice, the relationship between ionic mass and ICR frequency in Equation 1 is not sufficiently accurate to provide for "exact" mass determination of an unknown ion from the measured ICR frequency of a single known q/m peak. At least three major sources of frequency shifts are recognized: a shift due to the trapping potential; line shape distortion and frequency shift due to space-charge and coulomb repulsion

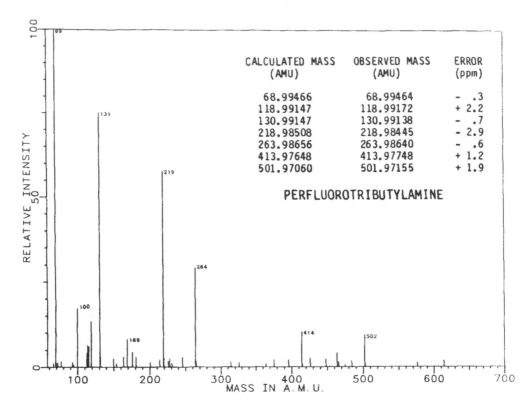

FIGURE 9.   Fourier transform mass spectral calibration over approximately 1 decade in mass, for perfluorotributylamine data obtained in approximately 0.1 sec using a Nicolet FTMS-1000 spectrometer. The masses
are for singly charged positive ions. (Spectrum courtesy of Dr. Chip Cody.)

between ions of like sign; and shifts due to nonuniform electrostatic potentials from
the various cell plates. Although progress in quantifying these effects has been made
(see, for example, References 19 and 20), the simplest approach is to fit several measured "known" peaks to an empirical equation of the form[21]

$$m = \frac{A}{\nu} + \frac{B}{\nu^2} \qquad (4)$$

in which ν is the observed ICR peak frequency. Exact mass determination for an unknown peak (by either interpolation or extrapolation from Equation 4) is then sufficiently accurate (<5 ppm) to determine the chemical formula in most cases. For example, a sample calibration table is shown in Figure 9 for perfluorotributylamine.

D. Low-Volatility Solid Samples

A major problem in mass spectrometry of solid ("probe") samples is that the sample
must be heated to increase its vapor pressure. However, since most mass spectrometers
require a sample pressure of 10^{-5} to 10^{-6} torr, heating the sample may result in pyrolysis
before the vapor pressure of the parent neutral is high enough to detect by electron
ionization. Because the FT-MS instrument routinely operates at a sample pressure
about 1000 times lower than that of a conventional mass spectrometer (e.g., 10^{-8} to
10^{-9} torr), the FT-MS technique is particularly well suited for use with low-volatility
samples. Figure 10 shows an FT-MS spectrum of an organometallic mixed-metal clus-
ter compound, obtained at 100°C at a sample pressure of about 2×10^{-8} torr. The

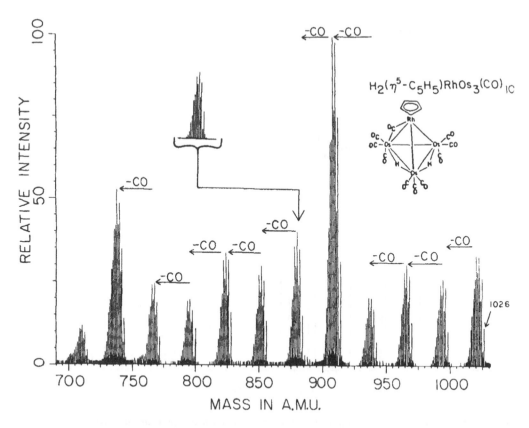

FIGURE 10. Fourier transform positive-ion mass spectrum of H$_2$(η^5-C$_5$H$_5$)RhOs$_3$(CO)$_{10}$ run as a probe sample at 100°C, with a Nicolet FTMS-1000 spectrometer. (From Hsu, L.-Y., Hsu, W.-L., Jan, D.-Y, Marshall, A. G., and Shore, S. G., *Organometallics*, 3, 591, 1984. With permission.)

parent ion and fragments corresponding to loss of each of the ten carbonyl groups are clearly visible.[18]

E. Mass Range

Unlike E/B double-focusing mass spectrometers, the upper mass limit of which at fixed accelerating voltage is determined by the maximum available magnetic field strength, an FT-MS spectrometer has no inherent upper mass limit. Sensitivity at fixed excitation amplitude is independent of mass.[6] Figure 11 shows a negative-ion spectrum of polystyrene 2350 in which masses exceeding 3700 are clearly evident. The spectrum was obtained by Dr. R. Cody (Nicolet Instrument Corporation) using laser desorption (see below).

F. Laser Desorption/Ionization

Figure 12 shows a positive-ion spectrum from daunorubicin (daunomycin), a close relative of the widely used antitumor drug doxorubicin (adriamycin). The parent mass is 527, and the base (i.e., largest) peak is for the (M + K)$^+$ ion at 566. The fragment at 437 corresponds to loss of the amino sugar moiety. Laser desorption has the advantage of producing a clean spectrum with a prominent pseudomolecular ion, from a relatively small amount of sample (say, 100 ng in 1 μl of solvent). The illustrated spectrum was obtained from a single laser pulse (i.e., no signal averaging). The spectrum was obtained by Dr. R. Cody from a sample provided by Dr. D. Horton (Ohio State University).

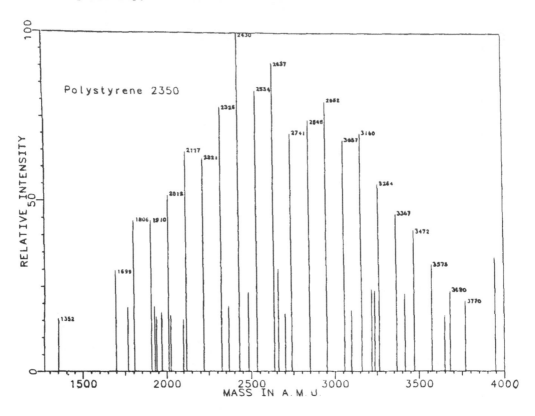

FIGURE 11.   Laser desorption Fourier transform positive-ion mass spectrum of polystyrene 2350 and gramicidin S, using a Tachisto model 215G pulsed CO_2 laser interfaced to a Nicolet FTMS-1000 spectrometer. (Spectrum courtesy of Dr. Chip Cody.)

G. Cesium Ion Desorption/Ionization

Cesium ion desorption/ionization has recently been adapted to FT-MS.[22] Although the detailed fragmentation pattern depends on the incident beam angle, it is clear that the method is capable of producing pseudomolecular ions at very high mass. For example, the spectrum in Figure 13 shows a cesium cationized dimer of vitamin B_{12} at m/z = 2792: $[(B_{12})_2 + Cs - 2CN]^+$.

H. Chemical Ionization

For the most typical mass spectral experiment (electron impact ionization of a solid sample), the molecular ion mass spectral peak is weak and may even be absent altogether. Figure 14 (top) shows an FT-ICR positive-ion mass spectrum of a deuterated phenobarbital sample. The parent ion at m/z = 237 is visible, but weak. As is often the case, the parent neutral molecule has a higher proton affinity than the daughter fragments. Therefore, if the ions are left to equilibrate for 1 sec, the protonated parent ion $(M + H)^+$ now becomes the largest peak in the spectrum, facilitating identification of the compound (Figure 14, bottom). For this "self chemical ionization" experiment, it is important to note that the spectrometer configuration is identical except for the additional 1-sec delay between ion formation and ion detection.

I. Gas-Phase Acidity and Basicity

A traditional application of ion cyclotron resonance has been the determination of gas-phase ion-molecule reaction rate constants and equilibrium constants, particularly gas-phase acidities and basicities.[23] The Fourier transform approach simply makes it

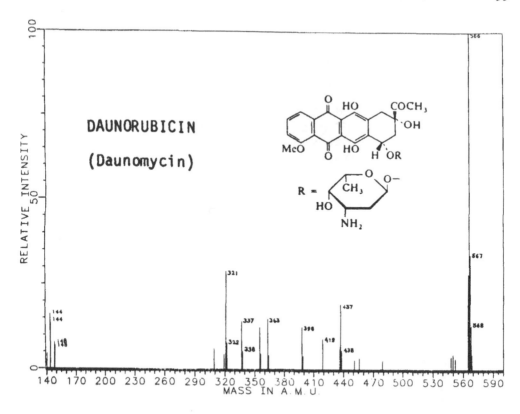

FIGURE 12. Laser desorption Fourier transform positive-ion mass spectrum of daunorubicin, obtained as in Figure 11.

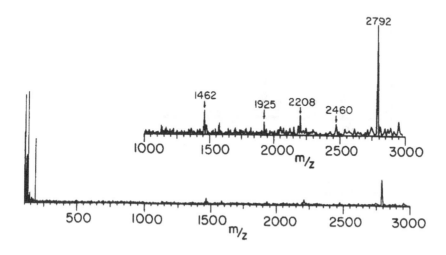

FIGURE 13. Fourier transform mass spectra obtained by Cs⁺ desorption of vitamin B_{12} from a gold wire grid. The superimposed spectrum is normalized to unit intensity at m/z = 2792. (From Castro, M. E. and Russell, D. H., *Anal. Chem.*, 56, 578. With permission.)

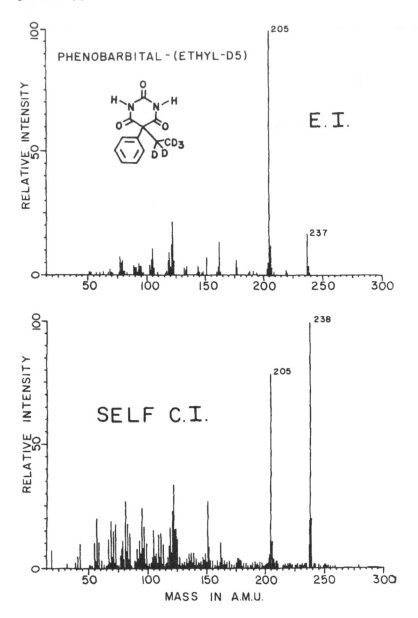

FIGURE 14.   Electron impact (top) and self chemical ionization (bottom) Fourier transform positive-ion mass spectra for deuterated phenobarbital. The two
spectra differ only by the addition of a 1-sec delay between ion formation and ion
excitation/detection (see text). Data were obtained from a Nicolet FTMS-1000
instrument.

possible to conduct those experiments much more easily: SNR, mass resolution, and
speed can be improved by factors of up to 100, 5000, and 10,000, respectively.

For example, a gas-phase basicity ($- \Delta G°$ for the reaction, $M + H^+ \rightleftharpoons MH^+$) can be
determined simply by mixing an unknown with a compound of known basicity, ioniz-
ing the sample, waiting about 10 sec for the ions to equilibrate, and measuring the
relative mass spectral peak heights of the two protonated parent ions. Figure 15 shows
such an experiment for (unknown) cubane mixed with (known) cyclohexanone. The
gas-phase basicity for cubane could then be determined to be 198.7 kcal/mol, based

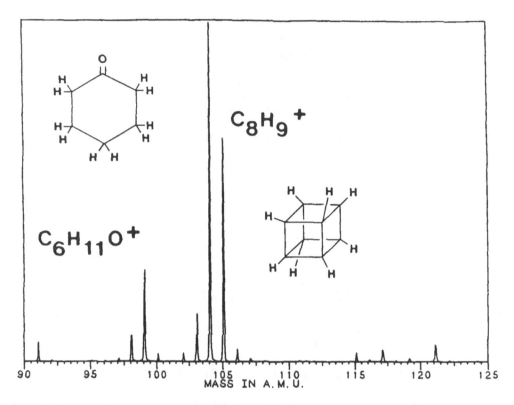

FIGURE 15. Positive-ion Fourier transform mass spectrum for a mixture of cubane and cyclohexanone at approximately equal partial pressures. Ions formed by electron impact were allowed to equilibrate for approximately 1 sec before this spectrum was taken. Cubane has the higher gas-phase basicity, as determined from the relative intensities of the two protonated parent ions ($C_8H_9^+$ and $C_6H_{11}O^+$). Data were obtained from a Nicolet FTMS-1000 instrument. (Cubane was provided by Dr. L. A. Paquette.)

on the measured partial pressure for the two parents and the known gas-phase basicity of cyclohexanone.

J. MS/MS/···

One of the most powerful recent mass spectral techniques is tandem mass spectrometry.[24,25] In its usual configuration, the first mass spectrometer filters out ions of all but one m/z value; the selected ions are then guided into a collision chamber, in which they are broken into fragments; finally, a second mass spectrometer sorts out the resultant fragment ions. In this way, it is possible to separate and identify two or more ions of the same m/z value in the original spectrum, according to their different fragmentation patterns. Unfortunately, the experiment is very expensive because of the need for two mass analyzers and a collision chamber.

With FT-MS, the same experiment can be done in a single sample chamber. After ions have been generated, ions of all but one m/z value are subjected to radiofrequency excitation powerful enough to eject them completely from the apparatus; the ions are then allowed to fragment collisionally (often with the aid of a collision gas introduced by a pulsed valve);[26] and finally, the resultant ion mass spectrum is detected in the usual way. Using such techniques, an MS/MS/MS/MS experiment has been performed with a single FT-MS spectrometer![27]

K. GC-FT-MS

A gas chromatograph can be coupled to an FT-MS instrument.[28] An example of a

Table 1
FT-MS ANALYTICAL
PERFORMANCE FEATURES

Sample pressure	$0.1-1000 \times 10^{-9}$ torr
Mass resolution	To 1,000,000 at m/z = 200
Speed	0.01—1 sec per scan
EI/CI/Laser/Cs$^+$	One source
MS/MS/$\cdots$	One spectrometer
+/− Ions	One knob
Upper mass	No geometric limit
Slits	None
High voltage	None
Automation	9 knobs/switches

gas chromatogram based on the integrated FT-MS spectral signal is shown for a human serum sample in Figure 16 (top). The peak eluting at 36 min is clearly identified as cholesterol from its mass spectrum in Figure 16 (bottom). The results of Figure 16 were obtained by Dr. J. Kinsinger (Nicolet Instrument Corporation) from a sample provided by Dr. B. Andresen. With FT-MS instruments now being developed, it should be possible to produce GC-FT-MS results with high chromatographic resolution *and* with mass resolution sufficiently high (about 20,000 at m/z 200) to yield exact mass determinations for individual GC peaks.

III. SUMMARY: ADVANTAGES, PROBLEMS, AND PROSPECTS

Some operating features of FT-MS are listed in Table 1. The speed of detection results from the Fellgett multiplex-advantage of detecting the entire spectrum at once,[5] rather than scanning from one end of the mass range to the other. Moreover, the instrument is structurally simple because ion formation, reaction, and detection all occur in a single sample chamber. Because of the absence of narrow slits and high voltages, and because of the low operating pressure, the instrument is extremely reliable — electron filaments can last for several months in normal operation. The instrument is simple to operate — virtually all adjustments and data manipulation are performed from a terminal keyboard.

One problem with the technique is that when too many ions (>100,000 or so) are present, coulomb repulsion leads to line broadening, and the space charge from the ions shifts the cylcotron frequency of the ions, making mass calibration and exact mass measurement more difficult. For most experiments, the number of ions can be kept low enough to avoid the problem, but the varying number of ions present during GC scans presents a serious problem for mass calibration. A second problem is the need for low pressure (<10^{-8} torr) in order to obtain ultrahigh mass resolution — this problem is being solved using fast and/or differential pumping. A third problem is the difficulty in obtaining accurate relative peak heights, due to nonlinear excitation amplitude across the mass spectrum;[11] use of stochastic excitation should solve the problem.[29]

FT-MS applications have recently been reviewed.[30,31] For newer ideas, the reader is referred to the collected abstracts of the 1985 American Association for Mass Spectrometry Meeting.[32] In summary, FT-MS offers the unique combination of ultrahigh mass resolution and speed. In addition, the technique is extremely versatile: exact mass, EI/CI/laser sources, MS/MS/$\cdots$, positive or negative ions, GC-MS, and multiple-ion monitoring can all be performed on the same instrument without hardware modifications.

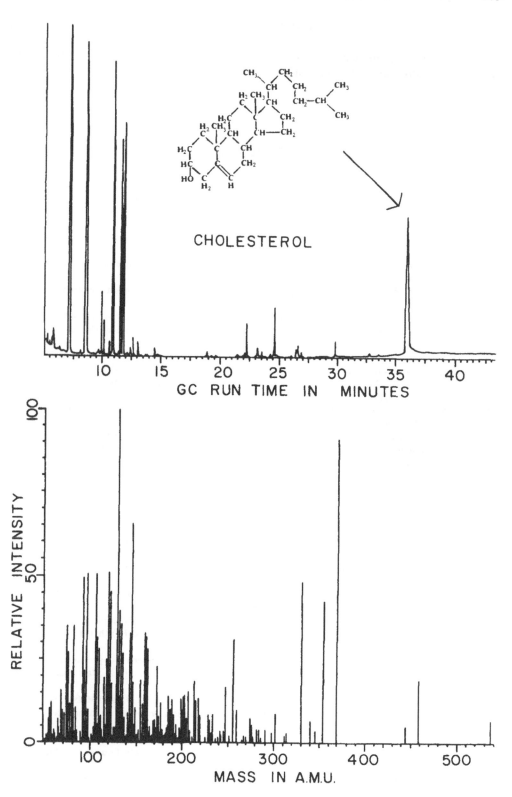

CHOLESTEROL

GC RUN TIME IN MINUTES

RELATIVE INTENSITY

MASS IN A.M.U.

FIGURE 16.   GC/FT/MS of a human serum sample. Top: total ion gas chromatogram; bottom: Fourier transform positive-ion mass spectrum (bottom) of the (cholesterol) peak eluting at 36 min.

ACKNOWLEDGMENTS

This work was supported by grants from the National Institues of Health (1 RO1 GM-31683), Sohio, and The Ohio State University.

REFERENCES

1. Marshall, A. G., Advantages of transform methods in chemistry, in *Fourier, Hadamard, and Hilbert Transforms in Chemistry,* Marshall, A. G., Ed., Plenum Press, New York, 1982, 1.
2. Marshall, A. G., Transform techniques in chemistry, in *Physical Methods of Modern Chemical Analysis,* Vol. 3, Kuwana, T., Ed., Academic Press, New York, 1983, 57.
3. Lehman, T. A. and Bursey, M. M., *Ion Cyclotron Resonance Spectrometry,* Interscience, New York, 1976.
4. Comisarow, M. B. and Marshall, A. G., Fourier transform ion cyclotron resonance spectroscopy, *Chem. Phys. Lett.,* 25, 282, 1974.
5. Comisarow, M. B. and Marshall, A. G., Frequency-sweep Fourier transform ion cyclotron resonance spectroscopy, *Chem. Phys. Lett.,* 26, 489, 1974.
6. Comisarow, M. B., Signal modeling for ion cyclotron resonance, *J. Chem. Phys.,* 69, 4097, 1978.
7. Comisarow, M. B. and Marshall, A. G., Selective-phase ion cyclotron resonance, *Can. J. Chem.,* 52, 1997, 1974.
8. Marshall, A. G., Comisarow, M. B., and Parisod, G., Relaxation and spectral line shape in fourier transform ion cyclotron resonance spectroscopy, *J. Chem. Phys.,* 71, 4434, 1979.
9. McIver, R. T., Jr., Trapped ion analyzer cell for ion cyclotron resonance spectroscopy, *Rev. Sci. Instrum.,* 41, 555, 1970.
10. Marshall, A. G., Convolution Fourier transform ion cyclotron resonance spectroscopy, *Chem. Phys. Lett.,* 63, 515, 1979.
11. Marshall, A. G. and Roe, D. C., Theory of Fourier transform ion cyclotron resonance mass spectroscopy: response to frequency-sweep excitation, *J. Chem. Phys.,* 73, 1581, 1980.
12. Comisarow, M. B. and Marshall, A. G., Theory of Fourier transform ion cyclotron resonance mass spectroscopy. I. Fundamental equations and low-pressure line shape, *J. Chem. Phys.,* 64, 110, 1976.
13. Comisarow, M. B. and Marshall, A. G., Resolution-enhanced Fourier transform ion cyclotron resonance spectroscopy, *J. Chem. Phys.,* 62, 293, 1975.
14. Comisarow, M. B. and Melka, J. D., Error estimates for zero-filling in Fourier transform spectrometry, *Anal. Chem.,* 51, 2198, 1979.
15. Marshall, A. G. and Comisarow, M. B., Multichannel methods in spectroscopy, in *Transform Techniques in Chemistry,* Griffiths, P. R., Ed., Plenum Press, New York, 1978, 39.
16. Marshall, A. G., Theoretical signal-to-noise ratio and mass resolution in Fourier transform ion cyclotron resonance mass spectrometry, *Anal. Chem.,* 51, 1710, 1979.
17. White, R. L., Ledford, E. B., Jr., Ghaderi, S., Wilkins, C. L., and Gross, M. L., Resolution and signal-to-noise in Fourier transform mass spectrometry, *Anal. Chem.,* 52, 1525, 1980.
18. Hsu, L.-Y., Hsu, W.-L., Jan, D.-Y., Marshall, A. G., and Shore, S. G., Synthesis and spectra of (^{5}-C_5H_5) $RhOs_2(CO)_9$ and $H_2(^5$-$C_5H_5)RhOs_3(CO)_{10}$; crystal structure of (^{5}-$C_5H_5)RhOs_2(CO)_9$, *Organometallics,* 3, 591, 1984.
19. Hunter, R. L., Sherman, M. G., and McIver, R. T., Jr., An elongated trapped-ion cell for ion cyclotron resonance mass spectrometry with a superconducting magnet, *Int. J. Mass Spectrom. Ion Processes,* 50, 259, 1983.
20. Francl, F. J., Sherman, M. G., Hunter, R. L., Locke, M. J., Bowers, W. D., and McIver, R. T., Jr., Experimental determination of the effects of space charge on ion cyclotron resonance frequencies, *Int. J. Mass Spectrom. Ion Processes,* 54, 189, 1983.
21. Ledford, E. B., Jr., Rempel, D. L., and Gross, M. L., *Anal Chem.,* 56, 2744, 1984.
22. Castro, M. E. and Russell, D. H., Cesium ion desorption ionization with Fourier transform mass spectrometry, *Anal. Chem.,* 56, 578.
23. Lias, S. G., Liebman, J. F., and Levin, R. D., An evaluated compilation of gas phase basicities and proton affinities of molecules: heats of formation of protonated molecules, *J. Phys. Chem. Ref. Data,* 13, 695, 1984.
24. McLafferty, F. W., Tandem mass spectrometry, *Science,* 214, 180, 1981.
25. Cooks, R. G. and Glish, G. L., Mass spectrometry/mass spectrometry, *Chem. Eng. New,* 30, 40, 1981.

26. Carlin, T. J. and Freiser, B. S., Pulsed valve addition of collision and reagent gased in Fourier transform mass spectrometry, *Anal. Chem.*, 55, 571, 1983.

27. Kleingeld, J. C., Ph.D. thesis, University of Amsterdam, The Netherlands, 1984.

28. Ledford, E. B., Jr., White, R. L., Ghaderi, S., Wilkins, C. L., and Gross, M. L., Coupling of capillary gas chromatograph and Fourier transform mass spectrometer, *Anal. Chem.*, 52, 2450, 1980.

29. Marshall, A. G., Wang, T.-C. L., and Ricca, T. L., Ion cyclotron resonance excitation/de-excitation: a basis for stochastic Fourier transform ion cyclotron mass spectrometry, *Chem. Phys. Lett.*, 105, 233, 1984.

30. Wanczek, K. P., *Int. J. Mass Spectrom. Ion Proc.*, 60, 11, 1984.

31. Gross, M. L. and Rempel, D. L., *Science (Washington, D.C.)*, 226, 261, 1984.

32. American Society for Mass Spectrometry, 33rd Annu. Conf. Mass Spectrometry and Allied Topics, San Diego, Calif., May 1985.

33. Comisarow, M. B., in *Fourier, Hadamard, and Hilbert Transforms in Chemistry*, Marshall, A. G., Ed., Plenum Press, New York, 1982, 133.

Chapter 5

ELECTRON SPECTROSCOPY FOR CHEMICAL ANALYSIS: APPLICATIONS IN THE BIOMEDICAL SCIENCES

Buddy D. Ratner and Brien J. McElroy

TABLE OF CONTENTS

I. INTRODUCTION

The number of new analytical tools available to the biomedical researcher is rapidly increasing; hence this volume. A frequently observed pathway for the development of these new techniques starts in the laboratories of physicists and physical chemists, moves after a period of instrument development into the laboratories of analytical chemists, and finally works its way into biological and biomedical laboratories. Electron spectroscopy for chemical analysis (ESCA, also called X-ray photoelectron spectroscopy, XPS) is in the transition region between the analytical chemists' laboratory and the widespread introduction into the biomedical research community. This review article will describe the surface analysis technique called ESCA, discuss its potential for application to biomedical problems, and review the small but growing literature on the application of ESCA to biological and biomedical problems.

Before formally considering specific details of the theory and practice of ESCA, it is worthwhile discussing why surface-analysis techniques are important in the biomedical sciences. The significance of surface analysis to the study of problems in biology, biomaterials, and medicine becomes clear if one realizes that many biologic processes take place in thin-film environments and at interfaces. Thus, protein molecules, cell walls, lipid membranes, and micelles are all of interest in the study of biological systems. The structures of these assemblies and the reactions that occur on them or within them are important to the proper functioning of biological systems. However, the small mass of material in these systems and the inability, in many cases, to separate the thin films participating in biological reactions from the bulk of attached tissue material or synthetic implant material make the techniques that study only the surface regions attractive.

Surface analysis methods have their primary utility in biomaterials science in five areas:

1. Unknown identification. The surface region of a solid is almost invariably different in composition from that in the bulk of the material. Two reasons for this are the enhanced surface chemical reactivity due to accessibility of a surface to reactants, and a thermodynamic driving force to minimize interfacial energy. In addition, for biomedical applications, surfaces are often consciously modified to achieve new properties (modifications to cell culture substrates with collagen or with sulfuric acid are good examples). Finally, manufacturers of biomedical devices are rarely as candid (or as cognizant) about the precise nature of their product as fundamental researchers would prefer. For these reasons, the ability to specifically analyze and identify the nature of a surface region is particularly important.

2. Contamination detection. Surfaces are readily contaminated with substances that are plentiful in our environment. Silicones and hydrocarbons are two common examples of materials that rapidly deposit upon surfaces. Surface analysis tools are useful, and often irreplaceable, since they allow the detection of the minute amounts of contaminant material that one often finds at interfaces and they provide a means to develop protocols to clean the surface or optimize preparative techniques to minimize contamination.[1] Surface analysis for contamination detection in the biomedical sciences has been reviewed.[2]

3. Reproducibility assurance. Since surfaces are so readily contaminated, one must always be concerned in biomedical studies about the ability to achieve the same surface from one day to the next. In addition, a manufacturer of a biomedical implant device must attempt to ensure that the device will perform identically each time a physician uses it. ESCA and other surface analysis tools can help to

achieve this important reproducibility assurance between experiments and between batches of devices or materials.

4. Quality control. If an affinity chromatography column is to be prepared, a cell culture substrate modified, a polymer film deposited on a glass surface, or any of a large variety of other preparations involving small amounts of material used to alter surfaces, one would like to have a direct (nonbiological) means to assay the quality of the preparation. Using surface analysis tools, questions such as "is the glass plate completely covered by the polymer?" or "has the correct amount of ligand been attached to the surface of the affinity column?" can be answered. Often, surface-analysis methods are the only tools available to unambiguously address these questions.

5. Surface property/biological response correlation. Biological assays for evaluating the responses of biological systems to various types of surfaces and interfaces are often expensive and time consuming. It would certainly be desirable to be able to perform a relatively straightforward physical measurement and get information that might hint at the biological reaction that could occur. The potential of surface analysis to predict biological responses and prescreen surfaces for performance in biological or medical applications has been suggested by a number of studies.[3-11]

There are many useful analytical tools that can be applied to the surface analysis of biomaterial systems. These include contact angle methods, attenuated total reflection (ATR) infrared spectroscopy, scanning electron microscopy (SEM), secondary ion mass spectrometry (SIMS), and Auger electron spectroscopy (AES). However, ESCA has the potential to provide more useful information about a biomaterials system than any of the other methods because of its high information content, its strong surface localization, and its advanced state of development. A complete surface analysis is best performed by obtaining corroborative and unique information utilizing a variety of surface-analysis methods. A number of review articles have appeared describing this multitechnique approach to surface analysis.[12-20] However, this chapter will deal only with the ESCA technique.

II. ESCA — THE TECHNIQUE

A. General Description

When matter is bombarded by electromagnetic radiation (photons), low-energy electrons are emitted by a process often called photoionization. The energy with which these electrons are ejected from the core orbital levels of atoms is related to the atomic and molecular environment from which they emerge. The ESCA experiment consists of measuring the numbers of electrons emitted at specific energies after low-energy X-ray bombardment. A plot of electron energy vs. number of electrons emitted is the usual manner in which ESCA data are collected and presented. Electron energy can be expressed either in terms of the kinetic energy (KE) actually measured in the ESCA instrument for the photoemitted electron or as the binding energy (BE) with which that electron was held within its parent atom. The relationship between the binding energy and the kinetic energy is

$$BE = h\nu - KE - \phi \tag{1}$$

where $h\nu$ is the characteristic energy of the X-ray line used for excitation and ϕ is the work function of the spectrometer (a parameter independently determined for each spectrometer and used as a constant).

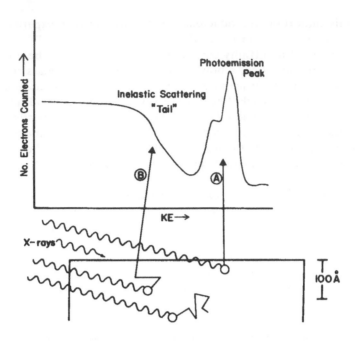

FIGURE 1.   X-ray-induced photoemission of electrons (A) from an atom near the surface can occur without interaction with other atoms. These electrons contribute to the photoemission peak. For atoms somewhat deeper into the specimen, interaction of emitted electrons (B) with other atoms has a greater probability of occurring. The loss of kinetic energy from such interaction results in the appearance of an inelastic scattering tail in the spectrum.

The X-rays used to induce the photoionization process have an excellent ability to penetrate into matter (typical penetration depths are of the order of microns). However, the photoelectrons emitted from the sample have a very short penetration distance in matter. The emitted photoelectrons, if they are involved in collisions with other atoms within the substance under observation, will lose a substantial portion of their kinetic energy. Those electrons that, even after collision, have sufficient energy to escape into the vacuum of the spectrometer will contribute to an elevated background signal in the observed spectrum often referred to as the inelastic scattering tail (see Figure 1). Only the electrons emitted from the surface region (perhaps the uppermost 10 to 20 atomic layers) that do not become involved in collisions will contribute to the photoemission signal that is used analytically. This phenomenon accounts for the utility of ESCA as a surface-analysis method.

Utilizing the binding energy, peak shape, and intensity of the ESCA signal, the following information about a specimen can be obtained:

1.   All elements present in the surface region (except H and He)
2.   Semiquantitative analysis of those elements found in the surface
3.   Molecular bonding environment and oxidation state for most atoms in a surface
4.   Differentiation of aromatic and aliphatic structures (in some cases)[21]
5.   Suggestions concerning the nature of bonding interactions at accessible interfaces
6.   Surface electrical properties and surface charging
7.   Nondestructive depth profiling (composition as a function of depth into the surface)
8.   Destructive depth profiling of inorganic materials using ion etching or RF plasma methods

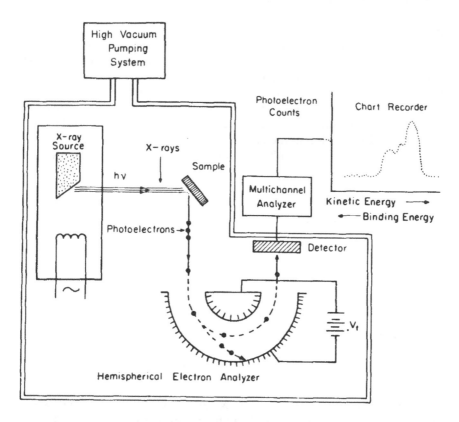

FIGURE 2. Schematic diagram of an ESCA system (From Ratner, B. D., *Ann. Biomed. Eng.*, 11, 313, 1983. With permission.)

9. Identification of specific organic functional groups based upon binding energy shifts and chemical derivatization studies
10. Fingerprint identification of materials by using valence band or molecular orbital spectra

A number of comprehensive volumes and review articles on the ESCA method have appeared and the reader is referred to them for more in-depth descriptions of the general method, instrumentation, and applications.[22-28]

The ESCA technique would appear to offer many advantages for biomedical studies. These include the large amount of information readily obtainable about a surface (see 1 to 10 above), the ability to work with samples of almost any shape and size in a nondestructive manner, the ability to study hydrated (frozen) specimens, and the preliminary studies that suggest that ESCA is observing parameters of importance in the regulation of reactions between surfaces and biological systems.

B. Instrumentation

The ESCA instrument must provide for X-ray excitation of a specimen and subsequent electron detection as a function of energy. This must be done in an ultrahigh vacuum and yet sample insertion must be relatively quick and easy. To achieve these ends, compromises in instrument design are made to balance energy resolution, spectral sensitivity, analysis speed, spatial resolution, and cost. The following description outlines the component parts that make up the most popular commercially available ESCA instrument designs. A schematic diagram of an ESCA system is presented in Figure 2.

1. X-Ray Source

The ESCA X-ray source must deliver a high-intensity monoenergetic photon flux to the analysis surface. The X-rays are generated by impinging a high-energy electron beam onto a target anode. Electron impact causes inner-shell vacancies which are filled by electrons from higher energy levels. An X-ray photon of a characteristic energy is subsequently emitted. There are several characteristic energy "lines" created in this way. In addition, a background radiation continuum, called Bremsstrahlung, is produced by electron deceleration in the target.

The ESCA peak width is the result of the convolution of the intrinsic atomic energy width, the X-ray line width, and instrumental broadening. The demand for narrow ESCA peaks then precludes many X-ray target materials because the energy widths are too large. The most common target materials are aluminum and magnesium. The aluminum $K_{\alpha_{1,2}}$ X-ray line has an energy of 1486.6 eV and a width of 0.85 eV. The magnesium $K_{\alpha_{1,2}}$ X-ray line has an energy of 1253.6 eV and a width of 0.7 eV. In conventional X-ray sources, magnesium affords the best energy resolution. Also, in such sources a thin window made of $\sim$5 μm of aluminum is used to attenuate unwanted X-ray satellites ($K_{\alpha_{3,4}}$, K_β) and the background radiation.

The X-ray monochromator crystal is used to remove the unwanted satellites and background radiation at the same time improving the width of the desired line. Typically, Al K_α X-rays are diffracted by a quartz crystal by Bragg's law:

$$\lambda_{Al K_{\alpha_{1,2}}} = 2d \sin\theta \tag{2}$$

where λ is the wavelength of the incident X-rays, d is the distance between crystal planes, and θ is the angle between the X-ray beam and the surface. Only X-rays satisfying this condition will be reflected in phase and go on to strike the sample. In principle, Al $K_{\alpha_{1,2}}$ line widths of 0.4 to 0.5 eV can be achieved, and a significant increase in signal-to-noise ratio (SNR) of the photoemission spectra will be observed by elimination of the Bremsstrahlung radiation and aluminum satellites. The penalty for this improved signal quality is an order-of-magnitude reduction in $Al_{K_{\alpha_{1,2}}}$ intensity and a significant increase in expense to purchase a monochromatic X-ray source. Both monochromatic and conventional X-ray sources are common on commercial ESCA instruments. However, for optimum energy resolution and the potential to achieve the best SNR, monochromators are essential.

A recent advance in ESCA instruments is the "small spot" X-ray source.[29] Previously, ESCA analysis areas ranged from 5 to 100 mm². With the introduction of the "small spot" X-ray optics, areas from 0.02 to 0.8 mm² can be analyzed while maintaining monochromated energy resolution.

2. Analyzers

Electron energy analyzers are used to measure the distribution of photoelectron energies. Various types based on magnetic and electrostatic principles have been used as ESCA spectrometers. The electrostatic dispersion analyzers have become the most accepted design. In these instruments, the electrons pass between two metal plates that are kept at different potentials so that an electron of a given energy E ± ΔE will pass through the analyzer and be detected. The energy E is referred to as the pass energy. ΔE refers to the energy window detected at a fixed analyzer potential and single electron energy. Analyzer resolution (R) is then defined as

$$R = \frac{\Delta E}{E} \tag{3}$$

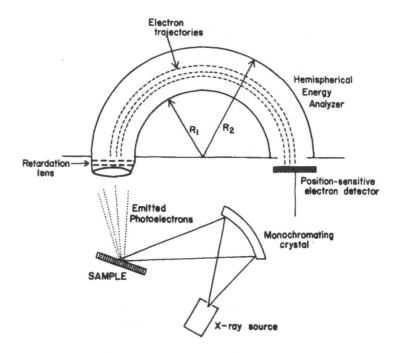

FIGURE 3. Schematic diagram showing an X-ray monochromater, hemispherical energy analyzer, and position-sensitive electron detector.

For a fixed analyzer resolution (R), lower pass energies yield a narrower energy window detected (ΔE). To take full advantage of this, instrument designers use a preretardation grid in front of the spectrometer to slow all the electrons to an operator-selected pass energy. The energy spectrum is acquired by sweeping the retardation voltage through the energy range desired. By maintaining a constant potential in the analyzer, a constant energy resolution through the energy spectrum is achieved. For a dispersive analyzer operating at a fixed pass energy, the overall efficiency[30,31] is given by

$$\text{eff} \; \alpha \; \delta E \int (I\Omega) \cdot dA \qquad (4)$$

where I is the photoelectron intensity/unit area/unit solid angle, A is the acceptance area into the analyzer, Ω is the solid angle of acceptance, and δE is the energy width "seen" by the detector. When a multichannel, position-sensitive detector is used to increase the effective energy window by simultaneously detecting several electron energies, the overall efficiency of the instrument goes up.

The hemispherical analyzer is the most popular design for ESCA spectrometers. A schematic diagram of a hemispherical analyzer is shown in Figure 3. The instrument depicted also has a monochromatic X-ray source and a position-sensitive detector.

Ideally, an electron entering the hemispherical analyzer will follow the path midway between the plates when it has a pass energy equal to E = eΔV, where e is the charge of an electron and ΔV is the potential difference across the plates given by

$$\Delta V = \frac{E}{e} \left(\frac{R_2}{R_1} - \frac{R_1}{R_2} \right) \qquad (5)$$

where the radii are defined in Figure 3. Electrons of slightly more or less energy will strike the detector at different locations. The position-sensitive detector is then used to

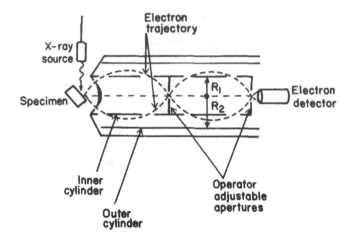

FIGURE 4.   Schematic diagram of a double-pass cylindrical mirror electron energy analyzer.

accumulate electron counts for several energies simultaneously. The increased data acquisition speed can partially compensate for the lower Al K_α X-ray intensity from the monochromator.

The two-stage cylindrical mirror analyzer (CMA) is shown in Figure 4. This design[32] can be used for ESCA and Auger electron spectroscopy (AES) interchangeably to take advantage of these complementary surface techniques. The CMA has been designed for high transmission of electrons to accommodate the AES requirements. The ESCA mode, electrons approaching the cylindrical analyzer with an angle of 42.3° ± 3° to the cylinder axis, will focus on the detector when the following relation is satisfied:

$$\Delta V = 0.76 \, \frac{E}{e} \, \ln\!\left(\frac{R_2}{R_1}\right) \tag{6}$$

where ΔV is the potential difference between the outer cylinder and the inner cylinder. A retardation grid is used to slow the entering electrons to a constant pass energy and constant resolution. The analyzer accepts a large solid angle with photoelectron take-off angles ranging from 40 to 155° making angle-dependent ESCA measurements extremely difficult. A conventional X-ray source is utilized and so spectral energy resolution and signal-to-noise ratios are lower than hemispherical analyzer systems, particularly if a monochromator is used on the hemispherical analyzer system. The CMA design also precludes the use of a position-sensitive detector.

3. Vacuum Pumping Systems

The ESCA spectrometer is enclosed in an ultrahigh vacuum (UHV) chamber operating in the 10^{-9} to 10^{-10} torr range. There are three conditions which require such low working pressures. First, X-ray sources, low-energy electron flood guns, accelerated ion sources, and mass spectrometers require low pressure to operate. Second, the likelihood of a photoelectron scattering with residual gas molecules before being detected must be reduced. A vacuum of 10^{-6} torr would adequately reduce the number of gas molecules and the subsequent scattering probability. The third consideration requires that the analysis surface not be contaminated by the vacuum system and that the surface composition remains constant during the course of an analysis. This restriction is critical because ESCA is highly surface sensitive. Even one contamination layer will strongly attenuate the substrate photoelectron signal. Oil-free cryogenic and ion pumps can be used to avoid the possibility of pump oil contamination. Cryogenic traps can minimize oil contamination from oil diffusion pumps.

To minimize contamination, very low pressures are needed. As an illustration, the kinetic theory of gases can be used to predict the number of molecules striking a unit area for a given temperature and pressure. Assuming, in the extreme limit, that each molecule that strikes the specimen sticks to it (sticking coefficient = 1), the contamination rate is approximately 1 monolayer per second at 10^{-6} torr, 1 monolayer per 15 min at 10^{-9} torr, and 1 monolayer per 2.5 hr at 10^{-10} torr. Fortunately, in practice, the sticking probability of most materials is less than unity. Polymer surfaces have the lowest sticking probabilities. Working pressures of 10^{-7} to 10^{-8} torr can be adequate for polymer systems.

To maintain a good vacuum in the ESCA spectrometers, special sample introduction chambers and insertion devices have been designed. For one commercial system, the introduction chamber is pumped with a roughing pump and a turbomolecular pump from atmospheric pressure to 10^{-4} torr. The sample, which is mounted on a carrier, is transferred to the main UHV chamber with an insertion rod that can be manipulated from outside the spectrometer. When the rod is removed and the main chamber, pumped using a cryopump, is isolated from the introduction chamber, low working pressures (10^{-8} torr) are quickly achieved.

Many types of vacuum pumps are available for ESCA and other UHV systems. Most commercial systems now use a roughing (rotary vane, rotary piston, or Roots) pump to assist in going from atmospheric pressure to 10^{-3} to 10^{-4} torr, and then various combinations of turbomolecular pumps, ion pumps, cryopumps, and titanium sublimation pumps to achieve high vacuum and UHV environments. Oil-diffusion pumps are less commonly used on contemporary instruments. Each of these pumping systems has advantages and disadvantages. Consequently, the choice of a pumping system will often depend upon the specific application and the instrument requirements for pump-down speed, minimized contamination, minimum required pressure, vibration isolation, and freedom from maintenance. Detailed descriptions of vacuum pumps, pressure measuring equipment, and vacuum systems have been presented.[33]

C. Chemical Shifts

The term "electron spectroscopy for chemical analysis" (ESCA) reflects the euphoria of Siegbahn and co-workers[34-36] when they discovered compound-related energy shifts of core-level photoelectron lines. This observation and related studies on ESCA as an analytical tool led to the awarding of the Nobel prize in physics to Dr. Siegbahn in 1981.[22] Examples of the chemical shifts observed in the carbon 1s (C1s) spectrum and oxygen 1s (O1s) spectrum of poly(ethylene terephthalate) are shown in Figure 5.

To help gain an intuitive understanding of the chemical shift, Siegbahn proposed a simplified model for interpreting chemical shifts based on analogy to the classical electrostatics problem of a hollow charged sphere.[34] The potential (V) inside the sphere is given by

$$\Delta V = \frac{\Delta q}{r} \tag{7}$$

where Δq is the change in charge density of the sphere and r is the radius. By assuming that the inner core levels do not interact with the valence "shell," then the change in binding energy of the inner-shell electrons due to a change in valence electron charge density or change in the average valence electron radius is

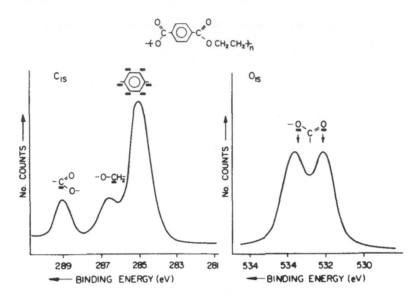

FIGURE 5.   Cls and Ols ESCA spectra of poly(ethylene terephthalate) showing chemical shifts associated with different carbon and oxygen molecular environments.

$$\Delta BE \; \alpha \; \frac{\Delta q}{r} \tag{8}$$

With this simple model, increased binding energy with increasing oxidation state and decreased interatomic distance is predicted.

The next level of sophistication in chemical shift models uses Koopmans' approximation.[37,38] It is assumed that the measured binding energy is approximated by the quantum energy levels in the neutral molecule. The chemically shifted binding energy (E_i) is then predicted by

$$E_i = E_{ref} + kq_i + \sum_j \frac{q_j}{R_{ij}} \tag{9}$$

where q_i is the charge on atom i, and q_j is the charge on atom i due to neighbors (j), and R_{ij} values are the interatomic distances. The values of k and E_{ref} are determined by the simultaneous equations of selfconsistent field theory. The equation itself is not particularly useful to ESCA users concerned with analytical problems. This model overlooks the fact that the measured binding energy reflects the energy difference between a ground state neutral atom and a singly ionized atom, and yet it predicts reasonably close values. The core levels of the neutral atom do not remain "frozen" during the ionization process. The orbitals undergo a reorientation or relaxation in the final ion state. The relaxation energy is usually opposite in sign to the predicted shift above and partially negates the error. Two other effects have been neglected above. The correlation energy is due to the pair-wise correlated motion of electrons in the atom. This energy must be considered when calculating the energy difference between an N electron atom and an (N-1) electron ion. The relativistic energy, which is greater for the most inner shells, is assumed the same for atom and ion. An excellent description of chemical shift theory is given by Fadley in Reference 30. Examples of the application of chemical shift information to the analysis of polymer surfaces are given in Section III. of this chapter.

D. Charging

A large number of electrons leave the sample as a result of X-ray bombardment. Conducting specimens in electrical contact with the spectrometer replenish the lost electrons and remain neutral by drawing a current from ground through the specimen holder. But for nonconducting specimens, a positive surface charge builds up as a result of the net loss of electrons. The resulting electrostatic potential slows escaping photoelectrons and the observed binding energy is higher by the magnitude of this potential.

Charging specimens have a floating binding-energy scale. The ESCA user is now faced with a calibration problem to be overcome before binding-energy shifts due to chemical bonding can be interpreted. One approach is to use the adventitious carbon contamination peak which appears on all but the most carefully prepared surfaces. The carbon layer is assumed to be in good "electrical contact" with the underlying substrate. A spectrum offset correction factor can be obtained by assigning to the advantitious carbon peak the binding energy 284.6 eV. This value is typically observed on thin carbon contamination layers on metal substrates. Caution should be used when interpreting published data on nonconductors because different values ranging from 284.6 to 285.0 eV have been used.

The best method of charge referencing is the use of an internal standard where the spectrum contains a peak of a known chemical shift and binding energy. In polymer systems the C—C peak at 285.0 eV is used since carbons in hydrocarbon-like molecular environments are often present in the polymer backbone.

In conjunction with charge correction of the binding-energy scale, a low-energy flood gun is often used to establish a dynamic equilibrium where the flood-gun electron supply rate is equal to the rate of electron loss thereby negating the charge shift. However, a perfect balance may not be optimum due to nonuniform surface charge build up.

Nonuniform surface charging can be encountered in dispersed multiphase materials (compositionally nonhomogenous surfaces) or in an ESCA spectrometer with a nonuniform X-ray flux. Observed peak broadening or even multiple peaks result from surface regions which have charge potentials greater than 0.2 eV apart. The use of a low-energy flood gun is essential to bring the ESCA analysis area to a uniform electrostatic potential and thereby reduce the observed peak width to that observed on homogenous samples.

For more detailed discussions of the use of a flood gun to deal with sample charging see References 39 and 40. Sample charging has also been dealt with by incorporating a UV source in the analysis chamber to generate a flux of low-energy electrons from the stainless-steel chamber walls.[41]

Another "external" charge calibration method relies on the direct deposition of gold onto the surface in submonolayer quantities. This method is generally not recommended because the gold segregates into islands on the surface,[42] and therefore it is not necessarily at a uniform electrostatic potential. Recently it has been observed that by carefully optimizing the amount of gold on a surface, reliable results can be obtained.[43]

E. Inelastic Mean Free Path

The surface sensitivity of ESCA arises from the short mean free path of low-energy electrons produced by low-energy X-ray bombardment. The mean free path (λ) is defined by the distance in which $1/e$ of the photoelectrons have escaped the surface without energy loss due to inelastic scattering. The underlying assumption carried with this definition is that the probability that a photoelectron will pass through the solid without losing energy by inelastic collision is given by:

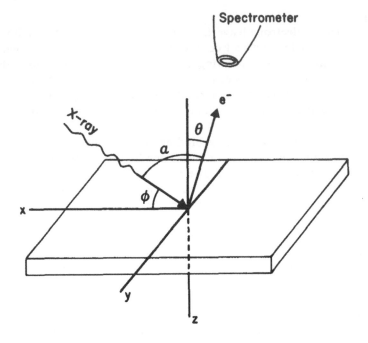

FIGURE 6. Geometric considerations in X-ray irradiation, photoemission, and electron detection.

$$\frac{N(Z)}{N_o(Z)} \propto \exp\left(\frac{-Z}{\lambda}\right) \tag{10}$$

where $N_o(Z)$ is the number of photoelectrons generated, $N(Z)$ is the number of photoelectrons that escape the solid without energy loss at depth Z (that is, with the original binding energy information intact), and Z is the distance from the surface (for simplicity the electron is assumed moving along the Z axis). The rationale for this expression can be seen when one integrates Equation 10 to an infinite depth for a homogenous material and obtains

$$\frac{N}{N_o} \propto \int_0^\infty \exp\left(\frac{-Z}{\lambda}\right) dZ \propto \lambda \tag{11}$$

a simple expression showing that the total fraction of photoelectrons leaving the sample is proportional to the mean free path. More will be said about the functional dependence of the mean free path later on.

Equations 10 and 11 must be modified when one has a spectrometer which detects electrons leaving the surface at an angle other than perpendicular to the surface. The geometry of the situation is shown in Figure 6 where the angle θ between the surface normal and the spectrometer is called the take-off angle. The mean free path (λ) is replaced by the escape depth ($\lambda_e = \lambda\cos\theta$) and Equation 11 is integrated along the axis of electron motion to obtain

$$\frac{N}{N_o} \propto \lambda_e = \lambda \cos\theta \tag{12}$$

where the fraction of detected photoelectrons can be adjusted by tilting the sample.

When the take-off angle is adjusted towards higher angles, the ESCA becomes even more sensitive to the outer surface layers.

How far into the surface does ESCA analyze? Because the number of photoelectrons arising from deeper layers and escaping the sample are functionally predicted by an exponential decay, there is an ambiguity in defining an exact ESCA analysis depth. For example when considering the depths $\lambda\cos\theta$, $2\lambda\cos\theta$ and $3\lambda\cos\theta$, the respective contribution to the photoelectron signal is 63.2, 86.5, and 95.0%. To avoid this impasse, the sampling depth is defined as $3\lambda\cos\theta$ and is frequently used to describe the ESCA analysis depth.

The inelastic mean free path has been studied both theoretically[44] and experimentally.[45-47] These efforts have shown a general dependence on electron energy and matrix effects. In the range of electron energies (150 to 1200 eV) typically encountered in ESCA, the mean free path is approximated by

$$\lambda \propto E^m \tag{13}$$

where m is an experimentally determined constant ranging in value from 0.5 to 1.5. Full agreement on the actual value of m has been hampered by the experimental difficulties in obtaining accurate measurements of λ from which m is determined. Perhaps the most useful approach to date to the general ESCA user is the expression for the mean free path developed by Seah and Dench.[47] They assumed a "universal" relation for all energies, $E^{0.5}$ power law to describe $E > 150$ eV, and E^{-2} power law to describe the dependence of electron scattering below 75 eV.

$$\lambda = \frac{A}{E^2} + B\,E^{0.5} \tag{14}$$

where A and B were determined by least-squares fit of large sets of published experimental data for each material class; namely, elements, inorganic compounds, and organic compounds.

The following expressions were obtained

$$\text{elements} \quad \lambda = \frac{538}{E^2} + 0.41(aE)^{1/2} \text{ (in monolayers)} \tag{15}$$

$$\text{inorganic compounds} \quad \lambda = \frac{2170}{E^2} + 0.72(aE)^{1/2} \text{ (in monolayers)} \tag{16}$$

$$\text{organic compounds} \quad \lambda = \frac{49}{E^2} + 0.11(E)^{1/2} \text{ (in mg/m}^2\text{)} \tag{17}$$

where a is the thickness of a monolayer in nanometers.[47]

The measurement of the inelastic mean free path in ESCA has been fraught with controversy (see, for example, References 48 and 49). However, the values of λ calculated using the above equations represent reasonable first approximations for most analytical work, particularly for organics. A few additional key references on the determination λ for organics are included in the reference list.[50-56] Further discussion of the inelastic mean free path concept used in conjunction with angular dependent ESCA data for nondestructive depth profiling will be presented shortly.

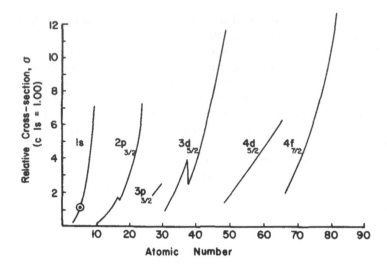

FIGURE 7.   Relative cross sections calculated for Al K, X-ray-induced photoemission.[58]

F. Quantitative Analysis

The starting point for quantitative calculations is the simple linear model stating that the observed intensity (I_i) of the photoemission signal is proportional to that obtained from a pure element (i) by the amount of fractional atomic concentration (X_i).[74]

$$I_i = X_i I_i^\infty \qquad (18)$$

I^∞ is the signal intensity for a surface consisting of 100% of component i. It is often difficult to generate the needed pure element data (I^∞) for this equation. Typically, published sets of reference data, expressed as I^∞/I^∞_{ref} are used. These ratios are referred to as relative sensitivity factors (S_i). The sensitivity factors are either empirically determined or calculated. When one has adjusted the sensitivity factor for any experimental parameters which differ from the published data set then the surface concentrations are determined by[74]

$$X_i = \frac{I_i/S_i}{\sum\limits^{n} I_j/S_j}$$

$$j = 1,2,3 \dots \qquad (19)$$

All that remains for quantitative calculation is to develop the correction terms to be incorporated into the sensitivity factor arising from the experimental details of the photoemission process and the ESCA instrument.[57]

Using the instrument geometry shown in Figure 6, the following experimental factors contribute to the relative intensities observed in ESCA. First, $J(h\nu)$, the X-ray intensity illuminating the specimen, is assumed to be constant across the surface and a function of X-ray energy. The X-rays excite the various subshells (k) of element (i) with differing probabilities, termed the cross section, $\sigma_{ik}(h\nu)$. Figure 7 shows the relative cross sections calculated by Scofield for aluminum K, X-rays.[58] The probability that a generated photoelectron will escape the surface decreases as $\exp[-Z/\lambda_i(E_k,M)\cos\theta]$ where $\lambda_i(E_k,M)$ is the inelastic mean free path of an electron in material matrix (M) and with a kinetic energy of (E_k).

The variation in photoemission intensity as the angle is varied between the X-ray source, sample surface, and analyzer must next be considered. Photoemission from ls' atomic subshells (zero angular momentum) is isotropic and symmetric. Thus, for a homogenous, clean specimen for which only 1s levels are being studied, one need not be concerned with a dependence of photoemission intensity with sample and instrument geometry. For this reason, studies on many clean, simple polymer surfaces for which only carbon 1s, oxygen 1s, nitrogen 1s, or fluorine 1s lines are observed can be performed on a number of different ESCA instruments and yield similar results. However, the situation where p, d, and f levels are being observed requires that the angular dependence of photoemission be considered. Photoemission from subshells of nonzero angular momentum is anisotropic. The intensity variation with angle of the photoelectrons, B, is proportional to

$$B(\beta_{ik}, \alpha) \propto 1 + \frac{\beta}{2} \left(\frac{3}{2} \sin^2 \alpha - 1 \right) \qquad (20)$$

where β_{ik} is one angular asymmetry parameter ($-1 < \beta_{ik} < 2$) for the element (i) and photoemission peak (k). The angle α is the angle between the incident X-ray beam and the escaping photoelectron. A published set of asymmetry parameters for commonly used X-ray sources is given in Reference 59. It should be noted that ESCA instruments may have different angles α due to differences in X-ray source mounting and would be expected to yield different relative peak intensities. The angular acceptance and efficiency of the electron spectrometer are collectively described as the transmission function $\Gamma(x,y,E_k)$ for each point in the analysis area. Imbedded within this function are experimental variables such as pass energy, aperture size, specimen alignment, and focusing. Depending on the instrument design, the transmission function can usually be estimated as a simple function of energy. However studies have shown that even among instruments of identical design there are deviations from the expected function and so it is important to empirically establish these data.[60]

Finally, for a selected pass energy, the electron detector will have a constant efficiency, D. The following relation between the observed intensity and the atomic density ϱ_i in the analysis volume can be expressed

$$I_i = \sigma_{ik}(h\nu) \ J(h\nu) \ D \ B(\beta_{ik}, a) \int_{-\infty}^{\infty} \int_{-\infty}^{\infty} \Gamma(x,y,E_k)$$

$$\int_{Z=0}^{\infty} \rho_i(xyz) \exp\left(-z/\lambda_i(E_k, M)\cos\theta\right) \, dx \, dy \, dz \qquad (21)$$

which can be simplified to

$$I_i = \sigma_{ik}(h\nu) \ J(H\nu) \ D \ B(\beta_{ik}, a) \cdot G(E_k) \cdot \lambda_{ik} \cos\theta \ N_i \qquad (22)$$

where G is the integrated transmission function and N_i is the number of atoms of element i. Now, when the pure element is referenced to a standard such as carbon (where $\sigma_{std} \equiv 1.0$) we obtain the relative sensitivity factor

$$S_i = \frac{I_i^\infty}{I_{std}^\infty} = \sigma_{ik} \frac{\lambda_{ik}}{\lambda_{std}} \frac{B(\beta_{ik})}{B(\beta_{std})} \frac{G(E_k)}{G(E_{std})} \qquad (23)$$

where X-ray intensity, detector efficiency, and takeoff angle(θ) are assumed to be the same for both measurements.

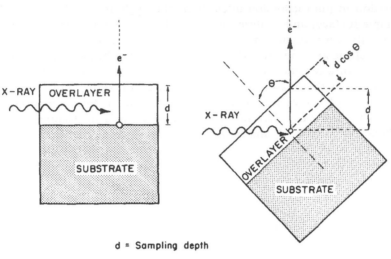

d = Sampling depth

d cos Θ = Effective Sampling thickness

FIGURE 8.   Decrease in the effective sampling depth as the angle of the specimen (θ) is increased with respect to the normal (X-ray source and detector in fixed positions). (From Ratner, B. D., Horbett, T. A., Shuttleworth, D., and Thomas, H. R., *J. Colloid Interface Sci.*, 83, 630, 1981. With permission.)

G. Angular Dependent Studies

The study of overlayers thinner than the ESCA sampling depth is an application at which ESCA excels. The ESCA technique allows one to determine the overlayer thickness and to examine the chemical state shifts at the substrate/overlayer interface. In addition, using special algorithms to treat the ESCA data, one can examine compositional gradients extending from the surface into the bulk. These depth-profiling analyses can be performed by ESCA in a nondestructive fashion using angular-dependent studies; i.e., studies in which the angle of the sample with respect to the horizontal is varied while keeping the position of the X-ray source and analyzer fixed.

For an overlayer thickness, d, of species, A, on a substrate of species, B, the ESCA intensity in Equation 18 can be modified so that attenuation of the substrate signal I_B due to inelastic scattering by overlayer A is taken into account.

$$I_B = I_B^{\infty} X_B \exp\left(-d/\lambda_B \cos\theta\right) \tag{24}$$

the corresponding equation for A is

$$I_A = I_A^{\infty} X_A \left[1 - \exp\left(-d/\lambda_A \cos\theta\right)\right] \tag{25}$$

As the takeoff angle θ is increased, the effective sampling depth is decreased and the signal I_A is enhanced relative to I_B (Figure 8).

These equations assume that the pure element reference spectra have been measured. They also assume that the specimen is smooth. Specimen roughness tends to limit the useful range of angle-dependent studies due to shadowing of the incident X-rays or escaping electrons and other geometric effects.[61,62]

A number of articles have been written dealing with methods for converting angular-dependent ESCA data into depth-profile information.[63-68] One such article in particular is of interest for polymeric systems since it allows the modeling of concentration gradients such as are most commonly found in the surface regions of polymers.[69] An example of an application of this algorithm to the angular-dependent ESCA data taken from a polyurethane surface is presented in Section III. of this chapter.

III. THE APPLICATION OF ESCA TO BIOMEDICAL PROBLEMS — POLYURETHANES

The way in which ESCA can be applied to biomedical problems will be illustrated using an actual example from the work of one of the authors (BDR). The system chosen for this example is that of polyurethanes, a class of polymers that show promise for use in heart-assist devices, blood vessel replacements, catheters, pacemaker lead insulation, and blood tubing.[70-73]

Polyurethanes of the type that are used in biomedical devices are usually synthesized in a two-stage process. First, a prepolymer is formed by the reaction of a diisocyanate with a low-molecular-weight (500 to 4000), dihydroxy-terminated polyether:

$$2(OCN\!-\!\!-NCO) + HO \sim\!\!\sim\!\!\sim OH \rightarrow OCN\!-\!\!\overset{\overset{\displaystyle O}{\|}}{N}CO \sim\!\!\sim\!\!\sim OCN\!-\!\!\overset{\overset{\displaystyle O}{\|}}{N}CO$$

    diisocyanate   polyether   prepolymer

Then this prepolymer is polymerized into the high-molecular-weight polyurethane using a reaction often referred to as chain extension:

$$OCN\!-\!\!\overset{\overset{\displaystyle O}{\|}}{\underset{\underset{\displaystyle H}{|}}{N}}CO \sim\!\!\sim\!\!\sim OCN\!-\!\!\overset{\overset{\displaystyle O}{\|}}{\underset{\underset{\displaystyle H}{|}}{N}}CO + NH_2CH_2CH_2NH_2 \rightarrow$$

$$\left(\!CN\!-\!\!NCO \sim\!\!\sim\!\!\sim OCN\!-\!\!NCNCH_2CH_2N\!\right)_n$$

poly(etherureaurethane)

The final polymer can have excellent mechanical properties, and materials in this class of polymers have been reported to have good tissue and blood compatibility. The need for surface analysis with polyurethanes occurs because, first, there are so many possible structures that can fall under the general category "polyurethane", and, second, there is evidence that the surface composition and bulk composition for polyurethanes differ widely. Since only the surface will contact or interface with the biological system of interest, we must understand the surface.

For a complete surface analysis of a biomedical polyurethane, the following ESCA experiments were performed:

1.   Wide-scan analysis to determine which elements were present in the polymer
2.   High-resolution analysis to determine the nature of the bonding environment for each element

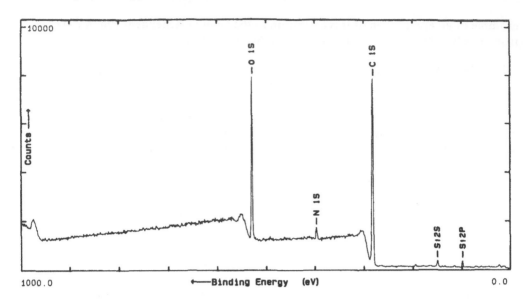

FIGURE 9.   Wide-scan ESCA spectrum from a polyurethane surface.

3.   Model compound studies to allow identification of organic functional groups present
4.   Extraction studies to understand surface stability
5.   Angular-dependent studies to look at surface compositional gradients
6.   Correlation between observed biological properties and ESCA parameters

In Figure 9 a wide-scan ESCA analysis of a polyurethane is shown (data acquired on a Surface Science Laboratories SSX-100 instrument). High-resolution spectra of most of the peaks in this plot are presented in Figures 10 through 13.* From the analysis over a 1000-eV range, all elements in the sample except hydrogen can be detected. The wide-scan spectrum, usually acquired at a relatively low-resolution (high count rate) instrument setting, provides the analyst with approximate peak binding energies so that the energy ranges that must be used to acquire each of the high-resolution scans are known. From the high-resolution scans, precise peak positions can be estimated and elemental concentration quantitation can be performed.

In order to identify the elemental core orbitals corresponding to each of the peaks on these plots, it is first necessary to correct the peak positions for shifts due to charging. Frequently, the carbon peak, which is almost always present, can be used as the reference peak for this charge correction. The carbon peak will generally be found in the area 270 to 290 eV (Figure 11). Let us assign the maxima for this peak to a generally accepted value for the organic carbon 1s electron binding energy, 285 eV. The basis for this assignment will be rationalized shortly when model compound studies are discussed. Once this assignment is made, one can subtract the observed C1s peak position from the assigned or standard peak position. The value obtained, the peak shift due to charging effects (6.6 eV in this case), can then be added to the observed binding energy for all other peaks obtained from this sample under these analysis conditions to give corrected binding energies. Figure 14 shows Figure 11 corrected for charging. The linear shift of the x axis is generally justified since in a properly calibrated ESCA instrument, there should be perfect linearity in binding energy over the entire binding energy range and the effect of charging on each core level line should be identical.

---

* Figures 9 through 13 are shown as taken from the ESCA instrument and are uncorrected for charging effects. This correction will be discussed.

125

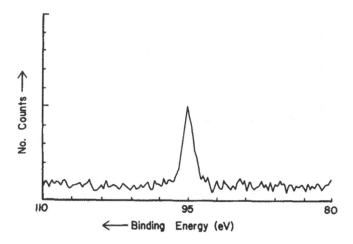

FIGURE 10. High-resolution ESCA spectrum of the silicon 2p region of the binding-energy spectrum.

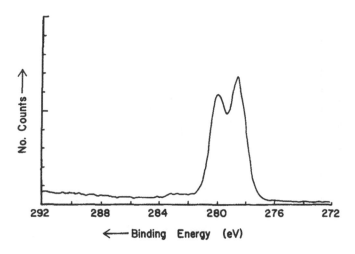

FIGURE 11. High-resolution ESCA spectrum of the carbon 1s region of the binding-energy spectrum.

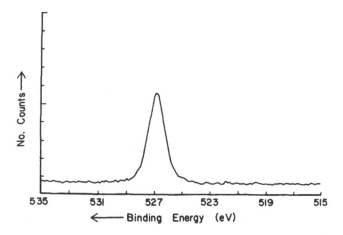

FIGURE 12. High-resolution ESCA spectrum of the oxygen 1s region of the binding-energy spectrum.

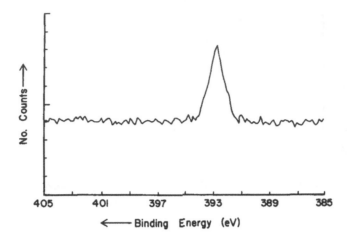

FIGURE 13.   High-resolution ESCA spectrum of the nitrogen ls region of the binding-energy spectrum.

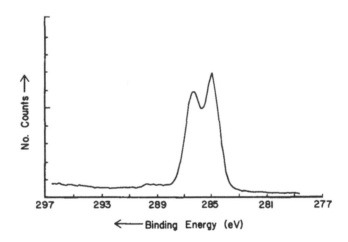

FIGURE 14.   The spectrum shown in Figure 11 is redrawn with the binding-energy scale shifted to correct for sample charging.

Once peak positions have been corrected, the elements with which they are associated can be more positively assigned using one of the many available tables of such data.[25,26,74] A condensed table of data on elemental lines commonly encountered in biomedical studies is presented as Table 1. Corroborative assignments (e.g., note the Si ls and Si 2p orbitals seen in Figure 9) that can often be made since more than one core orbital from some elements can be simultaneously observed, lend additional support to the element assignments.

Silicon, an easily detectable component of this polyurethane surface, is, of course, not a component of a polyetherurethane. However, in the form of organic silicone polymers, it is often found at the surface of materials and represents one of the most common contaminants observed. The other peaks seen in Figure 9 (C,O,N) are as expected for a polyurethane.

After assignment of the high-resolution peaks to appropriate elements, one can utilize the spectra to perform a quantitative analysis in the surface region. Such an analysis represents a complex average of the concentrations of the elements in the upper-

Table 1
ESCA BINDING ENERGIES AND
PHOTOEMISSION CROSS SECTIONS
FOR SOME LINES COMMONLY
OBSERVED IN BIOMEDICAL STUDIES

Line	Binding energy (eV)	Cross-section[a,58]
Cls	285	1.00
Nls	399	1.80
Ols	532	2.93
Fls	686	4.43
F2s	31	0.21
Nals	1072	8.52
Si2s	149	0.96
Si2p[b]	99/100	0.82
P2s	189	1.18
P2p[b]	135/136	1.19
S2s	229	1.43
S2p[b]	164/165	1.68
Cl2s	270	1.69
Cl2p	200/202	1.51/0.78
K2s	377	2.27
K2p	294/297	2.62/1.35
Ca2s	438	2.59
Ca2p	347/350	3.35/1.72

[a] Cross-sections given for aluminum $K_{\alpha_{1,2}}$ X-ray excitation.

[b] Typically observed as one peak, although actually a difficult to resolve doublet.

most approximately 50 Å of the sample. The accuracy of the quantitation is completely dependent upon the compositional homogeneity through the uppermost 50 Å region. If overlayers, lateral inhomogeneities, or concentration gradients exist at the surface, the meaning of the quantitation will not be straightforward.

Performing quantitative analysis begins by measuring areas under each of the photoemission peaks. These areas can be determined using computer software available with most ESCA systems or by cutting out and weighing the peaks. The area under the peaks is then divided by the number of instrumental scans that were used to accumulate the data. The peak area, normalized for number of scans, is next corrected for the differences in the efficiency with which each element emits electrons in response to X-ray excitation. Tables of such photoemission efficiency values, usually called photoemission cross sections or sensitivity factors, have been published.[28,58,74-76] Table 1, along with expected peak binding energies, also contains photoemission cross sections. In certain ESCA instruments (e.g., the HP5950 series) these values, when divided into the normalized peak areas, give numbers that, for a given vertically homogenous sample, are proportional to the surface concentration. For other instruments, additional correction must be made for the instrument transmission function (see the section of this article on quantitative analysis) and the energy dependence of the inelastic mean free path. Advice from the manufacturers of each specific instrument or experience with carefully chosen calibration standards is necessary to understand the required procedure for utilizing intensity data in a quantitative fashion. For the polyurethane shown in Figures 9 through 13, the peak areas and surface atomic percent composition are shown in Table 2. One important guideline in rigorously performing quantitative analysis using ESCA is that absolute peak intensity values cannot be compared between

Table 2

PEAK AREAS AND ATOMIC PERCENT COMPOSITION TAKEN
FROM THE POLYURETHANE DATA SHOWN IN FIGURES 9 TO 13

Line	Peak area	Scans (no.)	(Area/scan) = A_N	A_N/cross-section	Atomic %
Si2p	535	3	178	217	1.5
Cls	100335	8	12542	12542	89.0
Ols	25033	5	5006	1079	7.7
Nls	2740	6	457	254	1.8
					100.0

samples. Thus, for example, the %C or the C/N ratio for the specimen shown in this example can be compared to the %C or the C/N ratio for another polyurethane, but the number of counts under the Cls peak should not be directly compared with the number of Cls counts from another polyurethane specimen.

The peak shape seen in the high-resolution scan in Figure 14 is comprised of a number of subpeaks that overlap to varying degrees. The photoemission from the core atomic level of a single elemental species in one oxidation state or molecular environment will result in a symmetrical peak that is primarily Gaussian in character. By fitting a series of Gaussian peaks to the curve envelope in Figure 14, the contributions of each of the carbon-containing molecular species that make up the polyurethane can be estimated. There are three pieces of information needed to do this curve fitting in a meaningful fashion: the number of curves comprising the complex curve envelope, the widths of each of the subpeaks, and the peak positions that correspond to specific chemical functionalities.

The number of curves must usually be surmised from a preknowledge of the nature of the compound or from a rational guess based upon the width and appearance of the curve envelope. Such initial guesses are greatly aided by personal experience derived from studying a large number of compounds of known structure. Peak widths (all of which should be identical under a given curve envelope) and peak positions can be experimentally acquired from model compound studies.

Tables of peak shifts for a large number of organic functionalities and oxidation states are available.[25,28,77] However, if a specific class of compounds will be the focus of a major study, it is often preferable to study model compounds directly representative of functionalities or chemical states in that class of compounds. For a polyurethane study, model compounds indicative of each of the organic functional groups found in polyurethanes were studied.[78] Model compounds and their spectra are shown in Figure 15. In addition, Cls ESCA spectra from model polymers representative of the hard and soft segments in block copolyurethanes can be studied. Some information on expected peak shifts in polyurethanes derived from these model studies are presented in Table 3.

Once the number of peaks, their widths, and their binding-energy shifts are determined, analysis can proceed on, for example, the resolution of the spectrum in Figure 14 into its component peaks. Actual curve fitting can be performed using an analog curve resolver[79] or various computer Fourier transform, deconvolution, or least-squares curve-fitting routines that are available.[80-86] However, in all cases it is important to use constraints on the values of the peak width and the peak shifts. A computer will be able to perfectly fit any complex curve envelope with component peaks, but the results, if obtained without using constraints, although perhaps technically "the best fit," may have no physical significance in terms of known chemical species. The resolved Figure 14 is shown in Figure 16 along with the relative area fraction of each of the subpeaks. From this analysis, the fraction of hard and soft polyurethane segments in the surface region can be estimated, since the peak at 286.5 eV is primarily indicative

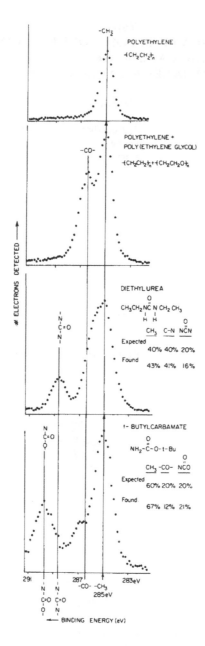

FIGURE 15. A model compound study to assign binding-energy shifts in the carbon 1s region to various functional groups commonly observed in polyurethanes. (From Ratner, B. D., in *Photon, Electron, and Ion Probes of Polymer Structure and Properties, ACS Symposium Series*, Vol. 162, Dwight, D. W., Fabish, T. J., and Thomas, H. R., Eds., American Chemical Society, Washington, D. C., 1981. With permission.)

Table 3
C1S ESCA PEAK POSITIONS FOR SOME
ORGANIC CARBON GROUPS FOUND IN
POLYURETHANES AND RELATED POLYMERS

Group	Description	Peak position (eV)
$-\underline{C}H$	Hydrocarbon	285.0
$-\underline{C}-N$	Amine	285.7—286.3
$-\underline{C}-O$	Ether, alcohol	286.5
$-\overset{\overset{O}{\|\|}}{\underline{C}}-$	Ketone	288.0
$-\overset{\overset{O}{\|\|}}{\underline{C}}-N-$	Amide	288.2
$-\overset{\overset{O}{\|\|}}{\underline{C}}-O-$	Acid, ester	288.8
$-N-\overset{\overset{O}{\|\|}}{\underline{C}}-N-$	Urea	288.8
$-N-\overset{\overset{O}{\|\|}}{\underline{C}}-O-$	Carbamate	289.3—289.7

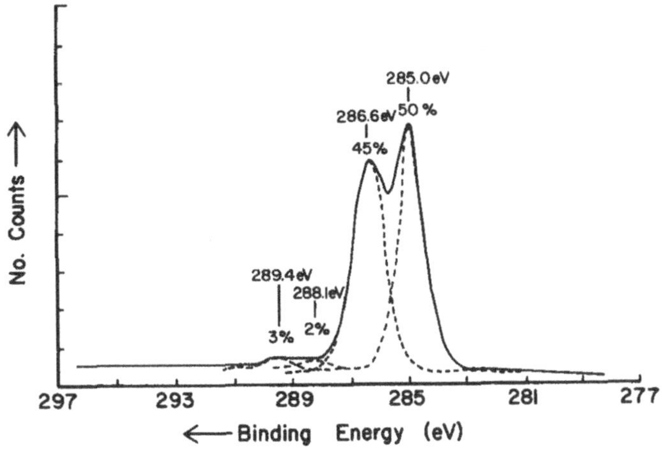

FIGURE 16.   The spectrum in Figure 14 resolved into its component
Gaussian peaks based on the model compound study.

of the polyether (soft segment) component and the carbamate peak at 289.4 eV is
related to the hard segment.

For an extruded polyurethane tubing (Pellethane® 2363-80A), a surface structure
that was not suggestive of a polyurethane was obtained.[87] In Figure 17a, amide functionalities, but not urethane groups, were detected. After extraction in solvents, the
surface structure was observed to be substantially different and indicative of a polyurethane (Figure 17b). This experiment allowed the identification of a stearamide extru-
sion lubrication at the surface of extruded Pellethane® and suggested a technique for
its removal.[87]

By performing an angular-dependent ESCA study on a smooth, cast polyurethane
film, a depth-profile model could be constructed based upon the data set obtained

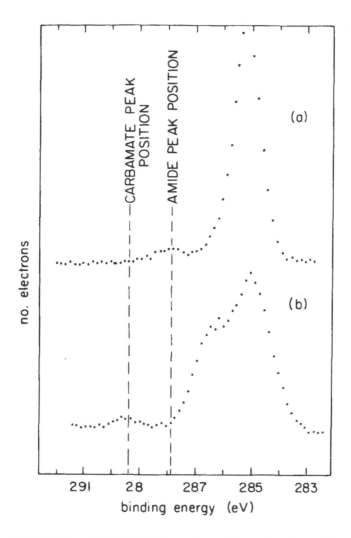

FIGURE 17. (a) The Cls ESCA spectrum of the surface of extruded Pellethane® tubing showing primarily hydrocarbon-like components and amide functionalities; (b) the surface of solvent-extracted, extruded Pellethane® tubing showing an ESCA Cls spectrum suggestive of a polyetherurethane. (From Ratner, B. D., in *Physiochemical Aspects of Polymer Surfaces*, Vol. 2, Mittal, K. L., Ed., Plenum Press, New York, 1983. With permission.)

(Figure 18; see Reference 88). This model suggests a concentration of polyether soft segment at the surface of the polyurethane.

Blood interaction data on a series of polyurethane tubes were taken using a baboon arteriovenous (AV) shunt model.[5] It was found that the fraction of the Cls peak in a hydrocarbon-like environment correlated well with the extent of platelet destruction observed for each type of polyurethane (Figure 19). Other studies have also observed correlations between ESCA spectra of polyurethanes and blood interactions.[3,11,89,90]

Section III. of this chapter was intended to illustrate the application of ESCA data to an active research area in biomaterials science, namely, polyurethanes. Although many ESCA techniques were brought to bear on this problem, a good deal more could still be learned. For example, the surface structure that exists at the hydrated polyurethane interface, certainly the relevant state for biomedical studies, could be examined by ESCA via the freeze-etch technique.[91] Using derivatization studies, more positive

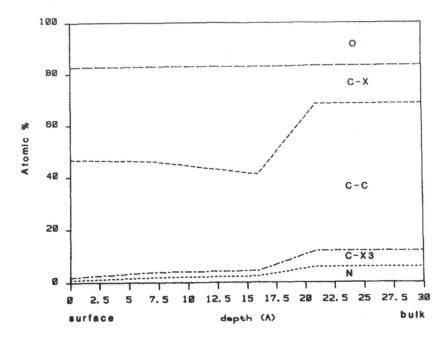

FIGURE 18.   A depth model based upon angular-dependent ESCA data taken from a polyurethane containing a mol wt 1000 poly(tetramethylene glycol) soft segment. C–C represents hydrocarbon-like functionalities resolved from the Cls spectrum, C–X represents carbon singly bonded to oxygen, and C–X₃ represents carbon species bound with two or more bonds to oxygen or nitrogen. (From Ratner, B. D., *Ann. Biomed. Eng.*, 11, 313, 1983. With permission.)

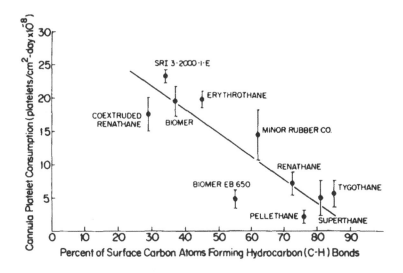

FIGURE 19.   The relationship observed between the fraction of the polyurethane Cls ESCA spectrum indicative of hydrocarbon-like environments and platelet consumption as measured in a baboon A-V shunt model. (From Hanson, S. R., Harker, L. A., Ratner, B. D., and Hoffman, A. S., in *Biomaterials 1980, Advances in Biomaterials*, Vol. 3, Winter, G. D., Gibbons, D. F., and Plenk, H., Jr., Eds., John Wiley & Sons, New York, 1982. With permission.)

assignments of the functional groups at the surface can be made.[92-97] Finally, using other corroborative surface analytical methods (contact angle measurements, SEM, and SIMS) a still more complete picture of the polyurethane surface can be pieced together.

IV. AN ANNOTATED BIBLIOGRAPHY OF THE BIOMEDICAL ESCA LITERATURE

Since commerical ESCA instrumentation first became available, a significant number of papers have appeared in which ESCA was applied to biomedical problems. Although the listing of papers presented here will not be complete and a thorough description of each will not be attempted, this annotated bibliography will allow the reader to get some perspective on the breadth of biomedical problems amenable to study by ESCA and the significance of the results that can be obtained. The bibliography will be arranged by important subject areas.

A. Protein and Biomolecule Studies

A casual search of the literature on the use of ESCA to study proteins turned up 34 references. Many dealt with the use of ESCA to look at metal-containing proteins.[98-103] Studies on protein derivatization have also been performed by ESCA.[104,105] Other biomolecules investigated by ESCA include fatty acids,[53,106,107] phospholipids,[108,109] amino acids,[110] and heparin.[111-113] ESCA has proven to be useful in the study of protein adsorption and binding to surfaces.[108,114,115] With appropriate data analysis, ESCA studies can yield information on the morphological organization of adsorbed protein films on surfaces.[116-118] Finally, there is at least one study that suggested that adsorbed protein films as studied by ESCA might be correlated with blood compatibility.[10] A recent review article dealing exclusively with ESCA studies of proteins addresses these issues in considerably more depth.[119]

B. Hydrogels

Hydrogels have been extensively investigated for biomedical applications.[120,121] Studies on their surfaces have been considered important for explaining the unique biological interactions exhibited by this class of materials. Surface-grafted hydrogels have been particularly attractive for study by ESCA because of the surface-localization of the graft and the small amounts of graft involved.[91,122-124] However, nongrafted hydrogel surfaces have also been examined.[95,114,123,125] ESCA would appear to be especially valuable for observing the compositional transitions that occur on hydrogel and other "mobile" surfaces during cycling between a hydrated and dehydrated condition.[91]

C. Polyurethanes

Polyurethane surfaces have been subject to extensive investigation. Section III. of this chapter discussed many aspects of the exploration of polyurethane surfaces by ESCA. Most articles dealing with ESCA studies of polyurethanes for biomedical applications address two issues: the surface-bulk compositional differentiation in polyurethanes[19,78,87,88,126-131] or the relationship between cell and blood interaction and polyurethane surface structure.[3,5,11,89,90,130,132-134] One study uses ESCA to assess chemical changes in a polyurethane surface after ion-beam texturing to make a porous surface for biomedical applications.[135]

D. Cell Culture Substrates

Small changes in the surface chemistry of cell culture substrates result in marked changes in their interactions with living cells. Since these surface chemistry changes take place in the upper few angstroms of material, ESCA becomes one of the few techniques that can effectively examine the chemistry of the reactions that occur. ESCA has been used to study polystyrene cell culture surfaces treated with H_2SO_4,[136,137] glass culture surfaces,[138] silane-treated culture surfaces,[139] treated poly(methyl methacrylate) surfaces,[140] and RF-plasma-reacted cell culture surfaces.[141-144] In one study, the surfaces of the cultured cells themselves were examined by ESCA.[145]

E. Dental and Orthopedic Applications

Relatively few publications have appeared describing the application of ESCA to materials problems in dentistry and orthopedics. This is surprising because bonding and interface problems are important to both fields and because metals and ceramics, commonly used materials in these fields, are both particularly amenable to study by ESCA. Dental-device bonding problems (porcelain-to-metal and polymer-to-enamel) have been studied in three publications.[146-148] The nature of tooth enamel and chemical interactions with the enamel have also been addressed.[149,150] Corrosion of metals used in orthopedic applications has been studied using ESCA after soaking in protein solutions[151] and in implant situations.[152] Finally, titanium surfaces have exhibited unique interfacial interactions with bone and have been the subject of preliminary ESCA investigations.[153]

V. CONCLUSIONS

In this chapter, we have attempted to present a comprehensible introduction to ESCA, particularly as it might affect the biological scientist who is interested in this technique to obtain supporting data. Much of the published literature on ESCA comes from the physics or physical chemistry communities and incorporates the jargon and research objectives from these fields. Consequently, practical and straightforward introductions to ESCA have been difficult to come by. We hope this chapter might assist with the understanding of the ESCA technique, while making no pretense at training expert ESCA spectroscopists. Realistically, most novices and casual ESCA users will need to work closely with experienced individuals since there is still much "art" and experience needed to work with the complex vacuum systems and electronics of the ESCA instruments to obtain artifact-free data, and to properly interpret the data. However, from this article, an appreciation can be developed for what is being done to samples submitted for analysis and how the data reduction is accomplished.

The need for ESCA studies of biomaterial surfaces and biological surfaces is clearly justified. The number of biological reactions which occur at interfaces and in thin-film environments, the ease with which surfaces are contaminated or altered, and the difficulties in using more traditional analytical methods to examine the minute amounts of material involved in interfacial processes support this statement. We hope this chapter will encourage the further development of ESCA in the biological sciences.

ACKNOWLEDGMENT

The authors acknowledge support of NIH grants RR01296 and HL25951 during the preparation of this manuscript and for some of the studies described herein.

REFERENCES

1. Ratner, B. D., Analysis of surface contaminants on intraocular lenses, *Arch. Ophthal.*, 101, 1434, 1983.
2. Ratner, B., Surface contamination and biomaterials, in *Treatise on Clean Surface Technology*, Mittal, K. L., Ed., Plenum Press, New York, 1985.
3. Sa Da Costa, V., Brier-Russel, D., Salzman, E. W., and Merrill, E. W., ESCA studies of polyurethanes: blood platelet activation in relation to surface composition, *J. Colloid Interface Sci.*, 80, 445, 1981.
4. Hanson, S. R., Harker, L. A., Ratner, B. D., and Hoffman, A. S., In vivo evaluation of artificial surfaces with a nonhuman primate model of arterial thrombosis, *J. Lab. Clin. Med.*, 95, 289, 1980.
5. Hanson, S. R., Harker, L. A., Ratner, B. D., and Hoffman, A. S., Evaluation of artificial surfaces using baboon arteriovenous shunt model, in *Biomaterials 1980, Advances in Biomaterials*, Vol. 3, Winter, G. D., Gibbons, D. F., and Plenk, H., Jr., Eds., John Wiley & Sons, New York, 1982, 519.
6. Chang, S. K., Hum, O. S., Moscarello, M. A., Neumann, A. W., Zingg, W., Leutheusser, M. J., and Ruegsegger, B., Platelet adhesion to solid surfaces. The effect of plasma proteins and substrate wettability, *Med. Prog. Technol.*, 5, 57, 1977.
7. Lyman, D. J., Muir, W. M., and Lee, I. J., The effect of chemical structure and surface properties of polymers on the coagulation of blood. I. Surface free energy effects, *Trans. Am. Soc. Artif. Intern. Organs*, 11, 301, 1965.
8. Yasuda, H., Yamanashi, B. S., and Devito, D. P., The rate of adhesion of melanoma cells onto nonionic polymer surfaces, *J. Biomed. Mater. Res.*, 12, 701, 1978.
9. Dexter, S. C., Sullivan, J. D., Jr., Williams, J., III, and Watson, S. W., Influence of substrate wettability on the attachment of marine bacteria to various surfaces, *Appl. Microbiol.*, 30, 298, 1975.
10. Fischer, J. P., Fuhge, P., Burg, K., and Heimburger, N., Methoden zur Herstellung und Charakterisierung von Kunststoffen mit verbesserter Blutvertraglichkeit, *Angew. Makromol. Chem.*, 105, 131, 1982.
11. Lelah, M. D., Lambrecht, L. K., Young, B. R., and Cooper, S. L., Physiochemical characterization and *in vivo* blood tolerability of cast and extruded Biomer, *J. Biomed. Mater. Res.*, 17, 1, 1983.
12. Ratner, B. D., Surface characterization of materials for blood contact applications, in *Biomaterials: Interfacial Phenomena and Applications, ACS Advances in Chemistry Series*, Vol. 199, Cooper, S. L. and Peppas, N. A., Eds., American Chemical Society, Washington D. C., 1982, 9.
13. Ratner, B. D., Characterization of graft polymers for biomedical applications, *J. Biomed. Mater. Res.*, 14, 665, 1980.
14. Morgan, A. E. and Werner, H. W., Modern methods for solid surface and thin film analysis, *Phys. Scr.*, 18, 451, 1978.
15. Holm, R. and Storp, S., Surface and interface analysis in polymer technology: a review, *Surf. Interface Anal.*, 2, 96, 1980.
16. Eisenberger, P. and Feldman, L. C., New approaches to surface structure determinations, *Science*, 214, 300, 1981.
17. Valenty, S. J., Example of problem solving using surface analysis, *J. Coat. Technol.*, 55, 29, 1983.
18. Turner, N. H., Dunlap, B. I., and Colton, R. J., Surface analysis: X-ray photoelectron spectroscopy, auger electron spectroscopy, and secondary ion mass spectrometry, *Anal. Chem.*, 56, 373R, 1984.
19. Graham, S. W. and Hercules, D. M., Surface spectroscopic studies of Biomer, *J. Biomed. Mater. Res.*, 15, 465, 1981.
20. Andrade, J. D., Surface analysis of materials for medical devices and diagnostic products, *Med. Dev. Diagnostics Ind.*, June 22, 1980.
21. Clark, D. T. and Dilks, A., ESCA studies of polymers. VII. Shake-up phenomena in some alkanestyrene copolymers, *J. Polym. Sci., Polym. Chem. Ed.*, 14, 533, 1976.
22. Siegbahn, K., Electron spectroscopy for atoms, molecules, and condensed matter, *Science*, 217, 111, 1982.
23. Dilks, A., X-ray photoelectron spectroscopy for the investigation of polymeric materials, in *Electron Spectroscopy: Theory, Techniques, and Applications*, Vol. 4, Baker, A. D. and Brundle, C. R., Eds., Academic Press, New York, 1981, 277.
24. Clark, D. T., Advances in ESCA applied to polymer characterization, *Pure Appl. Chem.*, 54, 415, 1982.
25. Carlson, T. A., *Photoelectron and Auger Spectroscopy*, Plenum Press, New York, 1975.
26. Briggs, D., *Handbook of X-ray and Ultraviolet Photoelectron Spectroscopy*, Heyden & Sons, London, 1977.
27. Ghosh, P. K., *Introduction to Photoelectron Spectroscopy*, John Wiley & Sons, New York, 1983.
28. Briggs, D. and Seah, M. P., *Practical Surface Analysis*, John Wiley & Sons, New York, 1983.
29. Kelly, M. A., More-powerful ESCA makes solving surface problems a lot easier, *Res. Dev.*, 26, 80, 1984.

30. Fadley, C. S., Basic concepts in x-ray photoelectron spectroscopy, in *Electron Spectroscopy: Theory, Techniques and Applications*, Vol. 2, Brundle, C. R. and Baker, A. D., Eds., Academic Press, New York, 1978, 2.

31. Fadley, C. S., X-ray photoelectron spectroscopy, in *X-ray Spectroscopy*, Azaroff, L. V., Ed., Mc-Graw-Hill, New York, 1974, 379.

32. Palmberg, P. W., A combined ESCA and Auger spectrometer, *J. Vac. Sci. Technol. B*, 12, 379, 1975.

33. O'Hanlon, J. F., *A User's Guide to Vacuum Technology*, John Wiley & Sons, New York, 1980.

34. Siegbahn, K., Nordling, C., Fahlman, A., Nordberg, R., Hamrin, K., Hedman, J., Johansson, G., Bergmark, T., Karlsson, S. E., Lindgren, I., and Lindberg, B., ESCA: atomic, molecular and solid state structure studied by means of electron spectroscopy, *Nova Acta Ragiae Soc. Sci. Ups., IV.*, 20, 5, 1967.

35. Hagstrom, S., Nordling, C., and Siegbahn, K., Electron spectroscopic determination of the chemical valence state, *Z. Phys.*, 178, 439, 1964.

36. Nordling, C., Hagstrom, S., and Siegbahn, K., Application of electron spectroscopy to chemical analysis, *Z. Phys.*, 178, 433, 1964.

37. Rabalais, J. W., *Principles of Ultraviolet Photoelectron Spectroscopy*, John Wiley & Sons, New York, 1977.

38. Koopmans, T. S., Über die Zuordnung von Wellenfunktionen und Eigenwerten zu den einzelnen Elecktronen eines Atoms, *Physica*, 1, 104, 1934.

39. Lewis, R. T. and Kelly, M. A., Binding-energy reference in x-ray photoelectron spectroscopy of insulators, *J. Electr. Spectrosc. Relat. Phenom.*, 20, 105, 1980.

40. Clark, D. T., Some experimental and theoretical aspects of structure, bonding and reactivity of organic and polymeric systems as revealed by ESCA, *Phys. Scr.*, 16, 307, 1977.

41. Dilks, A., Clark, D. T., Thomas, H. R., and Shuttleworth, D., ESCA applied to polymers. XXII. An investigation of sample charging phenomena for polymeric films on gold, *J. Polym. Sci. Polym. Chem. Ed.*, 17, 627, 1979.

42. Pashley, D. W., The nucleation, growth, structure and epitaxy of thin surface films, *Adv. Phys.*, 14, 327, 1965.

43. Uwamino, Y., Ishizuka, T., and Yamatera, H., Charge correction by gold deposition onto non-conducting samples in X-ray photoelectron spectroscopy, *J. Elect. Spectrosc. Relat. Phenom.*, 23, 55, 1981.

44. Penn, D. R., Quantitative chemical analysis by ESCA, *J. Elect. Spectrosc. Relat. Phenom.*, 9, 29, 1976.

45. Wagner, C. D., Davis, L. E., and Riggs, W. M., The energy dependence of the electron mean free path, *Surf. Interface Anal.*, 2, 53, 1980.

46. Powell, C. J., Attenuation lengths of low-energy electrons in solids, *Surf. Sci.*, 44, 29, 1974.

47. Seah, M. P. and Dench, W. A., Quantitative electron spectroscopy of surfaces: a standard data base for electron inelastic mean free paths in solids, *Surf. Interface Anal.*, 1, 2, 1979.

48. Clark, D. T., Thomas, H. R., and Shuttleworth, D., Electron mean free paths in polymers: a critique of the current state of the art, *J. Polym. Sci. Polym. Lett. Ed.*, 16, 465, 1978.

49. Cadman, P., Evans, S., Gossedge, G., and Thomas, J. M., Electron inelastic mean free paths in polymers: comments on the arguments of Clark and Thomas, *J. Polym. Sci. Polym. Lett. Ed.*, 16, 461, 1978.

50. Hupfer, B., Schupp, H., Andrade, J. D., and Ringsdorf, H., Photoelectron mean free paths in poly(diacetylene) mono- and multi-layers, *J. Elect. Spectrosc. Relat. Phenom.*, 23, 103, 1981.

51. Hall, S. M., Andrade, J. D., Ma, S. M., and King, R. N., Photoelectron mean free paths in barium stearate layers, *J. Elect. Spectrosc. Relat. Phenom.*, 17, 181, 1979.

52. Clark, D. T. and Thomas, H. R., Application of ESCA to polymer chemistry. XVI. Electron mean free paths as a function of kinetic energy in polymeric films determined by means of ESCA, *J. Polym. Sci. Polym. Chem. Ed.*, 15, 2843, 1977.

53. Brundle, C. R., Hopster, H., and Swalen, J. D., Electron mean-free path lengths through monolayers of cadmium arachidate, *J. Chem. Phys.*, 70, 5190, 1979.

54. Roberts, R. F., Allara, D. L., Pryde, C. A., Buchanan, D. N. E., and Hobbins, N. D., Mean free path for inelastic scattering of 1.2 keV electrons in thin poly(methyl methacrylate) films, *Surf. Interface Anal.*, 2, 5, 1980.

55. Lukas, J. and Jezek, B., Inelastic mean free paths of photoelectrons from polymer surfaces determined by the XPS method, *Coll. Czech. Chem. Commun.*, 48, 2909, 1983.

56. Clark, D. T., Abu-Shbak, M. M., and Brennan, W. J., Electron mean free paths as a function of kinetic energy: a substrate overlayer investigation of polyparaxylylene films on gold using a Ti K-alpha X-ray source, *J. Elect. Spectrosc. Relat. Phenom.*, 28, 11, 1982.

57. Seah, M. P., The quantitative analysis of surfaces by XPS: a review, *Surf. Interface Anal.*, 2, 222, 1980.

58. Scofield, J. H., Hartree-Slater subshell photoionizaion cross-sections at 1254 and 1487 eV, *J. Elect. Spectrosc. Relat. Phenom.* 8, 129, 1976.

59. Reilman, R. F., Msezane, A., and Manson, S. T., Relative intensities in photoelectron spectroscopy of atoms and molecules, *J. Elect. Spectrosc. Relat. Phenom.*, 8, 389, 1976.

60. Powell, C. J., Erickson, N. E., and Madey, T. E., Results of a joint Auger/ESCA round robin sponsored by ASTM committee E-42 on surface analysis. I. ESCA results, *J. Elect. Spectrosc. Relat. Phenom.*, 17, 361, 1979.

61. Ebel, M. F. and Wernisch, J., Shading at different take-off angles in X-ray photoelectron spectroscopy, *Surf. Interface Anal.*, 3, 191, 1981.

62. Wagner, N. and Brummer, O., The influence of various surface roughness on photoemission — (XPS-) parameters, *Exp. Tech. Phys.*, 29, 571, 1980.

63. Fadley, C. S., Solid-state and surface-analysis by means of angular-dependent x-ray photoelectron spectroscopy, *Prog. Solid State Chem.*, 11, 265, 1976.

64. Yamada, M. and Kuroda, H., Thickness measurement of surface layer by the angular dependent x-ray photoelectron spectroscopy, *Bull. Chem. Soc. Jpn.*, 53, 2159, 1980.

65. Ishizaka, A. and Iwata, S., Si-SiO$_2$ interface characterization from angular dependence of x-ray photoelectron spectra, *Appl. Phys. Lett.*, 36, 71, 1980.

66. Pijolat, M. and Hollinger, G., New depth-profiling method by angular-dependent x-ray photoelectron spectroscopy, *Surf. Sci.*, 105, 114, 1981.

67. Nefedov, V. I., X-ray photoelectron analysis of surface layers with composition gradients, *Surf. Interface Anal.*, 3, 72, 1981.

68. Ebel, M. F., Determination of the reduced thickness of surface layers by means of the substrate method, *J. Elect. Spectrosc. Relat. Phenom.*, 22, 157, 1981.

69. Paynter, R. W., Modification of the Beer-Lambert equation for application to concentration gradients, *Surf. Interface Anal.*, 3, 186, 1981.

70. Boretos, J. W., Past, present and future role of polyurethanes for surgical implants, *Pure Appl. Chem.*, 52, 1851, 1980.

71. Szycher, M. and Poirier, V. L., Synthetic polymers in artificial hearts: a progress report, *Ind. Eng. Chem. Prod. Res. Dev.*, 22, 588, 1983.

72. Lyman, D. J., New progress in biomedical polymer materials, *Pure Appl. Chem.*, 50, 427, 1978.

73. Lyman, D. J. and Knutson, K., Chemical, physical, and mechanical aspects of blood compatibility, in *Biomedical Polymers*, Goldberg, E. P. and Nakajima, A., Eds., Harcourt Brace Jovanovich, New York, 1980, 1.

74. Wagner, C. D., Riggs, W. M., Davis, L. E., and Moulder, J. F., *Handbook of X-Ray Photoelectron Spectroscopy*, Perkin-Elmer Corp., Eden Prairie, Minn., 1979.

75. Wagner, C. D., Davis, L. E., Zeller, M. V., Taylor, J. A., Raymond, R. H., and Gale, L. H., Empirical atomic sensitivity factors for chemical analysis by electron spectroscopy for chemical analysis, *Surf. Interface Anal.*, 3, 211, 1981.

76. Elliott, I., Doyle, C., and Andrade, J. D., Calculated core-level sensitivity factors for quantitative XPS using an HP5950B spectrometer, *J. Elect. Spectrosc. Relat. Phenom.*, 28, 303, 1983.

77. Clark, D. T. and Thomas, H. R., Applications of ESCA to polymer chemistry. XVII. Systematic investigation of the core levels of simple homopolyers, *J. Polym. Sci. Polym. Chem. Ed.*, 16, 791, 1978.

78. Ratner, B. D., ESCA and SEM studies on polyurethanes for biomedical applications, in *Photon, Electron, and Ion Probes of Polymer Structure and Properties*, ACS Symposium Series, Vol. 162, Dwight, D. W., Fabish, T. J., and Thomas, H. R., Eds., American Chemical Society, Washington, D. C., 1981, 371.

79. Clark, D. T. and Thomas, H. R., Applications of ESCA to polymer chemistry. X. Core and valence energy levels of a series of polyacrylates, *J. Polym. Sci. Polym. Chem. Ed.*, 14, 1671, 1976.

80. Koenig, M. F. and Grant, J. T., Deconvolution in x-ray photoelectron spectroscopy, *J. Elect. Spectrosc. Relat. Phenom.*, 33, 9, 1984.

81. Wertheim, G. K., Deconvolution and smoothing: applications in ESCA, *J. Elect. Spectrosc. Relat. Phenom.*, 6, 239, 1975.

82. Beatham, N. and Orchard, A. F., The application of Fourier transform techniques to the problem of deconvolution in photoelectron spectroscopy, *J. Elect. Spectrosc. Relat. Phenom.*, 9, 129, 1976.

83. Carley, A. F. and Joyner, R. W., The application of deconvolution methods in electron spectroscopy — a review, *J. Elect. Spectrosc. Relat. Phenom.*, 16, 1, 1979.

84. Verma, R., Profile deconvolution method for small computers, *Nucl. Instrum. Methods*, 212, 323, 1983.

85. Allen, J. D., Jr. and Grimm, F. A., High-resolution photoelectron spectroscopy through deconvolution, *Chem. Phys. Lett.*, 66, 72, 1979.

86. Dunn, W. L. and Dunn, T. S., An asymmetric model for XPS analysis, *Surf. Interface Anal.*, 4, 77, 1982.

87. Ratner, B. D., ESCA studies of extracted polyurethanes and polyurethane extracts: biomedical implications, in *Physiochemical Aspects of Polymer Surfaces*, Vol. 2, Mittal, K. L., Ed., Plenum Press, New York, 1983, 969.
88. Paynter, R. W., Ratner, B. D., and Thomas, H. R., An XPS and SEM study of polyurethane surfaces: experimental considerations, in *Polymers as Biomaterials*, Shalaby, S. W., Hoffman, A. S., Horbett, T. A., and Ratner, B. D., Eds., Plenum Press, New York, 1984.
89. Merrill, E. W., Salzman, E. W., Wan, S., Mahmud, N., Kushner, L., Lindon, J. N., and Curme, J., Platelet-compatible hydrophilic segmented polyurethanes from polyethylene glycols and cyclohexane diisocyanate, *Trans. Am. Soc. Artif. Intern. Organs*, 28, 482, 1982.
90. Lyman, D. J., Knutson, K., McNeill, B., and Shibatani, K., The effects of chemical structure and surface properties of synthetic polymers on the coagulation of blood. IV. The relation between polymer morphology and protein adsorption, *Trans. Am. Soc. Artif. Intern. Organs*, 21, 49, 1975.
91. Ratner, B. D., Weathersby, P. K., Hoffman, A. S., Kelly, M. A., and Scharpen, L. H., Radiation-grafted hydrogels for biomaterial applications as studied by the ESCA technique, *J. Appl. Polym. Sci.*, 22, 643, 1978.
92. Batich, C. D. and Wendt, R. C., Chemical labels to distinguish surface functional groups using X-ray photoelectron spectroscopy (ESCA), in *Photon, Electron and Ion Probes of Polymer Structure and Properties*, ACS Symposium Series, Vol. 162, Dwight, D. W., Fabish, T. J., and Thomas, H. R., Eds., American Chemical Society, Washington, D. C., 1981, 221.
93. Briggs, D., XPS studies of polymer surface modifications and adhesion mechanisms, *J. Adhesion*, 13, 287, 1982.
94. Briggs, D. and Kendall, C. R., Derivatization of discharge-treated LDPE: an extension of XPS analysis and a probe of specific interactions in adhesion, *Int. J. Adhesion Adhesives*, 2, 13, 1982.
95. Dickie, R. A., Hammond, J. S., De Vries, J. E., and Holubka, J. W., Surface derivatization of hydroxyl functional acrylic copolymers for characterization by x-ray photoelectron spectroscopy, *Anal. Chem.*, 54, 2045, 1982.
96. Everhart, D. S. and Reilley, C. N., Chemical derivatization in electron spectroscopy for chemical analysis of surface functional groups introduced on low-density polyethylene film, *Anal. Chem.*, 53, 665, 1981.
97. Pennings, J. F. M. and Bosman, B., Analysis of copolymer surfaces by surface reactions and XPS, *Colloid Polym. Sci.*, 258, 1099, 1980.
98. Weser, U., Jung, G., Ottnad, M., and Bohnenkamp, W., X-ray photoelectron spectroscopy (XPS) of bovine erythrocuprein, *FEBS Lett.*, 25, 346, 1972.
99. Walton, R. A., X-ray photoelectron spectra of copper complexes considered as models for metalloproteins containing copper-sulfur bonds, *Inorg. Chem.*, 19, 1100, 1980.
100. Leibfritz, D., X-ray photoelectron spectroscopy of iron-containing proteins — the valency of iron in ferredoxins, *Angew. Chem. Int.*, 11, 232, 1972.
101. Millard, M. M., X-ray photoelectron spectroscopic studies of biological materials: metal ion protein binding and other analytical applications, in *Protein-Metal Interactions*, Friedman, M., Ed., Plenum Press, New York, 1974, 589.
102. Kramer, L. N. and Klein, M. P., XPS of non-heme iron proteins, in *Electron Spectroscopy*, Shirley, D. A., Ed., North Holland, Amsterdam, 1972, 733.
103. Chiu, D., Tappel, A. L., and Millard, M. M., Improved procedure for X-ray photoelectron spectroscopy of selenium-glutathione peroxidase and application to the rat liver enzyme, *Arch. Biochem. Biophys.*, 184, 209, 1977.
104. Millard, M. M. and Friedman, M., X-ray photoelectron spectroscopy of BSA and ethyl vinyl sulfone modified BSA, *Biochem. Biophys. Res. Commun.*, 70, 445, 1976.
105. Millard, M. M. and Masri, M. S., Detection and analysis of protein functional group modification by X-ray photoelectron spectrometry, *Anal. Chem.*, 46, 1820, 1974.
106. Brown, C. A., Burns, F. C., Knoll, W., Swaien, J. D., and Fischer, A., Unusual monolayer behavior of a geminally disubstituted fatty acid. Characterization via surface plasmons and X-ray photoelectron spectroscopy study, *J. Phys. Chem.*, 87, 3616, 1983.
107. Clark, D. T. and Fok, Y. C. T., Electron mean free paths in Langmuir-Blodgett multilayers, *J. Elect. Spectrosc. Relat. Phenom.*, 22, 173, 1981.
108. Jaurand, M. C., Baillif, P., Thomassin, J. H., Magne, L., and Touray, J. C., X-Ray photoelectron spectroscopy and chemical study of the adsorption of biological molecules on chrysotile asbestos surface, *J. Colloid Interface Sci.*, 95, 1, 1995.
109. Baillif, P. and Touray, J. C., Etude de l'adsorption de monocouche a l'interface solide-solution diluee par spectroscopie de photoelectrons (XPS): application a l'adsorption de phospholipides sur des fibres d'amiante, *J. Elect. Spectrosc. Relat. Phenom.*, 32, 223, 1983.
110. Colton, R. J., Murday, J. S., Wyatt, J. R., and DeCorpo, J. J., Combined XPS and SIMS study of amino acid overlayers, *Surf. Sci.*, 84, 235, 1979.

111. Ebert, C. D. and Kim, S. W., Immobilized heparin: spacer arm effects on biological interactions, *Thromb. Res.*, 26, 43, 1982.

112. Ferruti, P., Barbucci, R., Danzo, N., Torrisi, A., Puglisi, O., Pignataro, S., and Spartano, P., Preparation and ESCA characterization of poly(vinyl chloride) surface-grafted with heparin-complexing poly(amido amine) chains, *Biomaterials*, 3, 33, 1982.

113. Lautier, A., Baquey, C., Salesse, A., Fougnot, C., Labarre, D., Basse-Cathalinat, B., Ducassou, D., and Josefowicz, M., ESCA studies on heparin-like materials used for extracorporeal shunts, *J. Artif. Intern. Organs*, 5, 199, 1982.

114. Gilding, D. K., Paynter, R. W., and Castle, J. E., Quantitative evaluation of water structuring and protein adsorption on the surface of hydrophilic polymers by ESCA, *Biomaterials*, 1, 163, 1980.

115. Sipehia, R. and Chawla, A. S., Albuminated polymer surfaces for biomedical application, *Biomater. Med. Devices Artif. Organs*, 10, 229, 1982.

116. Ratner, B. D., Paynter, R. W., Horbett, T. A., and Thomas, H. R., Adsorbed protein films investigated by ESCA, *Trans. Soc. Biomater.*, 6, 120, 1983.

117. Paynter, R. W., Ratner, B. D., Horbett, T. A., and Thomas, H. R., XPS studies on the organization of adsorbed protein films on fluoropolymers, *J. Colloid Interface Sci.*, 1984, in press.

118. Ratner, B. D., Horbett, T. A., Shuttleworth, D., and Thomas H. R., Analysis of the organization of protein films on solid surfaces by ESCA, *J. Colloid Interface Sci.*, 83, 630, 1981.

119. Paynter, R. W. and Ratner, B. D., The study of interfacial proteins and biomolecules by X-ray photoelectron spectroscopy, in *Surface and Interfacial Aspects of Biomedical Polymers*, Vol. 2, Andrade, J. D., Ed., Plenum Press, New York, 1985.

120. Ratner, B. D. and Hoffman, A. S., Synthetic hydrogels for biomedical applications, in *Hydrogels for Medical and Related Applications, ACS Symposium Series*, Vol. 31, Andrade, J. D., Ed., American Chemical Society, Washington, D. C., 1976, 1.

121. Ratner, B. D., Biomedical applications of hydrogels: review and critical appraisal, in *Biocompatibility of Clinical Implant Materials*, Vol. 2, Williams, D. F., Ed., CRC Press, Boca Raton, Fla., 1981, 145.

122. Tazuke, S. and Kimura, H., Surface photografting. II. Modification of polypropylene film surface by graft polymerization of acrylamide, *Makromol. Chem.*, 179, 2603, 1978.

123. Ratner, B. D. and Hoffman, A. S., Surface characterization of hydrophilic-hydrophobic copolymer model systems. I. A preliminary study, in *Adhesion and Adsorption of Polymers*, Part B, Lee, L., H., Ed., Plenum Press, New York, 1980, 691.

124. Hoffman, A. S., Horbett, T. A., Ratner, B. D., Hanson, S. R., Harker, L., A., and Reynolds, L. O., Thrombotic events on grafted polyacrylamide-Silastic surfaces as studied in a baboon, in *Biomaterials: Interfacial Phenomena and Applications, ACS Advances in Chemistry Series*, Vol. 199, Cooper, S. L. and Peppas, N. A., Eds., American Chemical Society, Washington, D. C., 1982, 59.

125. Miller, D. R. and Peppas, N. A., Surface analysis of poly(vinyl-alcohol-co-n-vinyl-2-pyrrolidone) by ESCA, *ACS Polym. Prepr.*, 24, 12, 1983.

126. Sung, C. S. P. and Hu, C. B., ESCA studies of surface chemical composition of segmented polyurethanes, *J. Biomed. Mater. Res.*, 13, 161, 1979.

127. Sha'aban, A. K., McCartney, S., Patel, N., Yilgor, I., Riffle, J. S., Dwight, D. W., and McGrath, J. E., ESCA studies on polydimethylsiloxane-urethane copolymers and their blends with segmented polyether-urethanes, *ACS Polym. Prepr.*, 24, 130, 1983.

128. Zhang, X., Pan C., and Qi, Z., ESCA studies on segmented polyurethanes, *Gaofenzi Tongxun*, 1982, 349, 1982.

129. Knutson, K. and Lyman, D. J., The effect of polyether segment molecular weight on the bulk and surface morphologies of copolyether-urethane-ureas, in *Biomaterials: Interfacial Phenomena and Applications, ACS Advances in Chemistry Series*, Vol. 199, Cooper, S. L. and Peppas, N. A., Eds., American Chemical Society, Washington, D. C., 1982, 109.

130. Nyilas, E. and Ward, R. S., Jr., Production of biomedical polymers. II. Processing technology/product property correlations of hemocompatible Avcothane elastomers, in *Science and Technology of Polymer Processing*, Suh, N. P. and Sung, N. H., Eds., MIT Press, Cambridge, Mass., 1979, 770.

131. Paynter, R. W., Ratner, B. D., and Thomas, H. R., Polyurethane surfaces — an XPS study, *ACS Polym. Prepr.*, 24, 13, 1983.

132. Sa Da Costa, V., Brier-Russell, D., Trudel, G., Waugh, D. F., Salzman, E. W., and Merrill, E. W., Polyether-polyurethane surfaces: thrombin adsorption, platelet adsorption, and ESCA scanning, *J. Colloid Interface Sci.*, 76, 594, 1980.

133. Lelah, M. D., Stafford, R. J., Lambrecht, L. K., Young, B. R., and Cooper, S. L., Cast vs. extruded biomer for biomedical applications, *Trans. Am. Soc. Artif. Int. Organs*, 27, 504, 1981.

134. Lyman, D. J., Albo, D., Jackson, R., and Knutson, K., Development of small diameter vascular prostheses, *Trans. Am. Soc. Artif. Int. Organs*, 23, 253, 1977.

135. Dwight, D. W. and Beck, B. R., Ion beam polymer surface modification: characterization by ESCA and SEM, *ACS Org. Coat. Plast. Chem. Prepr.*, 40, 494, 1979.

136. Maroudas, N. G., Sulphonated polystyrene as an optimal substratum for the adhesion and spreading of mesenchymal cells in monovalent and divalent saline solutions, *J. Cell. Physiol.*, 90, 511, 1977.
137. Curtis, A. S. G., Forrester, J. V., Mc Innes, C., and Lawrie, F., Adhesion of cells to polystyrene surfaces, *J. Cell Biol.*, 97, 1500, 1983.
138. Ratner, B. D., Rosen, J. J., Hoffman, A. S., and Scharpen, L. H., An ESCA study of surface contaminants on glass substrates for cell adhesion, in *Surface Contamination*, Vol. 2, Mittal, K. L., Ed., Plenum Press, New York, 1979, 669.
139. Ratner, B. D. and Swalec, J. J., Cell adhesion substrates prepared by vapor phase deposition of silane coatings, Abstracts, Fourth Annu. Mtg. Adhesion Soc., Savannah, Ga., February 23, 1981.
140. Yamanaka, A., Oshima, K., Ohira, A., Matsumoto, T., Nakamae, K., Furusawa, M., Goto, H., Takakura, K., Kawai, K., Yamamoto, S., and Moriuchi, T., Biocompatibility of plastics for intra-ocular lenses, *Afro-Asian J. Ophthalmol.*, 1, 47, 1982.
141. Andrade J. D., Iwamoto, G. K., McNeill, B., and King, R. N., XPS studies of polymer surfaces for biomedical applications, *ACS Org. Coat. Plast. Chem. Prepr.*, 36, 161, 1976.
142. Ratner, B. D., Haque, Y., Horbett, T. A., Schway, M. B., and Hoffman, A. S., RF plasma-deposited films as model substrates for studying biointeractions, 2nd World Congr. Biomaterials 10th Annu. Mtg. Soc. Biomaterials, Washington, D. C., April 27, 1984.
143. Andrade, J. D., Iwamoto, G. K., and McNeill, B., XPS studies of polymer surfaces for biomedical applications, in *Characterization of Metal and Polymer Surfaces*, Vol. 2, Lee, L. H., Ed., Academic Press, New York, 1977, 133.
144. Becton Dickinson, Primeria Tissue Culture Labware Product Brochure.
145. Millard, M. M. and Bartholomew, J. C., Surface studies of mammalian cells grown in culture by x-ray photoelectron spectroscopy, *Anal. Chem.*, 49, 1290, 1977.
146. Anusavice, K. J. J., Ringle, R. D., and Fairhurst, C. W., Identification of fracture zones in porcelain-veneered-to-metal bond test specimens by ESCA analysis, *J. Prosthet. Dent.*, 42, 417, 1979.
147. Ohno, H., Kanzawa, Y., and Yamane, Y., ESCA study on the oxidized surface of a gold alloy for porcelain-metal bonding, *Dent. Mater. J.*, 2, 59, 1983.
148. Bowen, R. L., Adhesive bonding of various materials to hard tooth tissues. XIV. Enamel mordant selection assisted by ESCA (XPS), *J. Dent. Res.*, 57, 551, 1978.
149. Hercules, D. M., Quantitative surface analysis using ESCA, *Phys. Scr.*, 16, 169, 1977.
150. Duschner, H., Uchtmann, H., and Ahrens, G., Electron spectroscopic determination of the calcium, phosphorus, oxygen, and fluorine ratios in ultra-thin layers of the enamel surface, *Dtsch. Zahnaerztl. Z.*, 35, 306, 1980.
151. Wortman, R. S., Merritt, K., Brown, S. A., and Millard, M. M., XPS analysis of metal salts, corrosion products, and protein interactions, *Trans. Soc. Biomater.*, 6, 105, 1983.
152. Hoffman, J., Michel, R., Holm, R., and Zilkens, J., Corrosion behaviour of stainless steel implants in biological media, *Surf. Interface Anal.*, 3, 110, 1981.
153. Albrektsson, T., Hansson, H. A., Kasemo, B., Larsson, K., Lundstrom, I., McQueen, D. H., and Skalak, R., The interface zone of inorganic implants *in vivo*: titanium implants in bone, *Ann. Biomed. Eng.*, 11, 1, 1983.
154. Ratner, B. D., Surface characterization of biomaterials by electron spectroscopy for chemical analysis, *Ann. Biomed. Eng.*, 11, 313, 1983.

Chapter 6

BIOLOGICALLY RELEVANT APPLICATIONS OF VIBRATIONAL CIRCULAR DICHROISM

Laurence A. Nafie

TABLE OF CONTENTS

I. INTRODUCTION

Vibrational circular dichroism is a relatively new member in the class of spectroscopic phenomena known as optical activity.[1-3] As such, it represents a new approach to the study of molecular structure. The primary purpose of this chapter is to introduce the reader to this new form of molecular spectroscopy. Along the way we will indicate some of the directions that applications have taken to date and offer a glimpse of how vibrational circular dichroism may likely have an impact on the study of biological systems in the years to come.

A. Optical Activity

The phenomenological occurrence of optical activity in nature has been known for more than a century.[2-4] In the absence of a magnetic field, optical activity is supported in a molecular sample only if the constituent molecules lack mirror symmetry. Under these circumstances the molecules are said to be chiral, or optically active. Optical activity can be defined in a general sense as the differential interaction of a molecule with left vs. right circularly polarized radiation. Of the various forms of optical activity, two are particularly well known — optical rotatory dispersion (ORD) and circular dichroism (CD).[5] ORD is defined as the spectrum of the angle of rotation of a plane of polarized light that is passed through a sample. The rotation arises from a difference in the index of refraction for light having left vs. right circular polarization, i.e., $n_L \neq n_R$, where the angle of rotation is proportional to $\Delta n = n_L - n_R$. CD on the other hand is defined as the spectrum of $\Delta\varepsilon = \varepsilon_L - \varepsilon_R$ where ε_L and ε_R are the absorption coefficients for left and right circularly polarized light, respectively. It can be demonstrated that the complete CD spectrum and the complete ORD spectrum for a molecule represent redundant sets of spectral information that are exactly related to one another by reciprocal integral transforms known as the Kronig-Kramers transformation.[6]

Until recently all manifestations of optical activity originated in transitions between different elecronic states in molecules. ORD was originally the preferred experimental quantity since rotations can be measured in all parts of the spectrum whereas CD can only be observed directly within an absorption band. On the other hand, CD is much simpler to interpret since the measurement of $\Delta\varepsilon$ at a given wavelength is related only to the transition properties of the molecule at that wavelength; the ORD value is more difficult to interpret since one must obtain information regarding transition properties that exist throughout the entire spectrum. Recent improvements in CD instrumentation have elevated the measurement of CD to favored status over that of ORD. The principal criterion for the observation of CD in a chiral molecule is that it must possess an electronic transition in an accessible part of the visible or ultraviolet (UV) region. Except for special instruments this means a transition between 185 and 700 nm.

Optical activity in electronic transitions has proven to be a valuable source of stereochemical information for a wide variety of biologically relevant molecules.[5] Proteins, small peptides, nucleic acids, transition metal complexes of biomolecules, and drug molecules have been studied extensively in CD. In most cases CD is used to monitor changes in the molecular conformation where it has been shown to be much more sensitive than ordinary absorption spectroscopy. The detailed interpretation of CD spectra in these samples has proven difficult due in part to the structural complexity of biologically relevant molecules and due to the lack of spectral resolution encountered in their electronic CD spectra.

B. Vibrational Optical Activity

Optical activity in vibrational transitions has evolved over the past decade.[1] From the spectra obtained to date it is quite clear that vibrational optical activity (VOA) contains a wealth of sensitive stereochemical information. On the other hand, VOA intensities are generally small in magnitude and therefore difficult to measure. In fact, experimental limitations prevented successful measurements of VOA for several decades, despite numerous attempts. VOA spectroscopy can be regarded as either a new area of application for optical activity or as a refinement of vibrational spectroscopy. It combines the stereosensitivity of optical activity with the functional specificity and local character of vibrational spectroscopy.

Two forms of VOA have been discovered and developed to date: vibrational circular dichroism (VCD), the extension of CD measurement into the infrared vibrational region of spectrum, and the analogous measurement in Raman spectroscopy, known as Raman optical activity (ROA). In both VCD and ROA, the polarization of the incident beam of radiation is modulated between left and right circular states and the difference in the intensities for these two states is recorded. In contrast to CD and ORD, VCD and ROA are not redundant sets of spectral data. Rather, VCD and ROA are complementary in nature, providing different views of the nuclear and electronic dynamics of a vibrating chiral molecule. The remainder of this review will focus on VCD and its applications. The interested reader is referred to Reference 1 and citations therein for further discussion of ROA.

II. VIBRATIONAL CIRCULAR DICHROISM

A. Theoretical Basis

The source of intensity in VCD spectra is the imaginary part of the vector dot product of the electric dipole transition moment with the magnetic dipole transition moment. For a transition between vibrational levels v and v′ of the ath normal mode, this quantity is called the rotational strength and is written as

$$R^{\alpha}_{v'v} = \mathrm{Im}(\mu^{\alpha}_{v'v} \cdot m^{\alpha}_{v'v})$$

The corresponding expression for ordinary infrared absorption intensity, the dipole strength, is written as

$$D^{\alpha}_{v'v} = |\mu^{\alpha}_{v'v}|^2$$

The transition moments are the matrix elements of the electric and magnetic dipole moment operators given by

$$\mu^{\alpha}_{v'v} = <v'_{\alpha}|\mu|v_{\alpha}> = <v'_{\alpha}|\sum_i e_i r_i|v_{\alpha}>$$

$$m^{\alpha}_{v'v} = <v'_{\alpha}|m|v_{\alpha}> = <v'_{\alpha}|\sum_i \frac{e_i}{2m_i c} r_i x p_i|v_{\alpha}>$$

where the operators are sums over all particles in the molecule and where e_i, m_i, r_i, and p_i are the charge, mass, position, and momentum of those particles (electrons and nuclei), and c is the speed of light. VCD can be viewed as arising from an interference between the electric dipole and magnetic dipole moment mechanisms for the absorption of infrared radiation. The combined dependence of VCD on the magnetic and electric dipole moments is what provides VCD spectra with additional sensitivity to stereochemical features. The rotational strength is simultaneously sensitive to the direction of the linear oscillations of charge (electric moments) and the sense of any circular oscillations of charge (which create magnetic moments).

The strength of VCD transitions is measured by the anisotropy ratio given by

$$g^{\alpha}_{v'v} = 4R^{\alpha}_{v'v}/D^{\alpha}_{v'v}$$

This ratio is a dimensionless quantity which is proportional to the ease of measurability of a VCD transition. It can also be directly related to the result of a VCD measurement by the following relation:

$$g^{\alpha}_{v'v} = \Delta\epsilon/\epsilon = \Delta A/A$$

Here $\Delta\epsilon = \epsilon_L - \epsilon_R$ and $\epsilon = (\epsilon_L + \epsilon_R)/2$ are measures of the VCD and absorption in terms of the molar absorptivity, ϵ. VCD may also be measured in terms of the decadic absorption, A, where $\Delta A = A_L - A_R$ and $A = \epsilon \ell C$ with pathlength, ℓ, and concentration, C.

B. Measurement Techniques

VCD intensities are four to five orders of magnitude smaller than ordinary infrared absorption intensities. The accurate measurement of VCD intensities to date has required a high-frequency polarization modulation scheme that possesses three critical requirements: (1) a polarization modulator that will create alternately left and right circular polarization in the infrared beam; (2) a low-noise infrared detector capable of following the polarization modulation; and (3) a low-noise, stable, phase-sensitive electronic detection system.[7,8] In a dispersive VCD spectrometer, infrared radiation is first mechanically chopped at low frequency ($\sim$100 Hz), dispersed through a grating monochromator, polarization modulated between left and right circular states at a frequency of $\sim$50 kHz using a photoelastic modulator constructed of CaF_2 or ZnSe, passed through the sample, and focused on a cooled semiconducting detector element such as InSb or HgCdTe. Dispersive VCD instruments operate best in the region above 2000 cm^{-1}; however, successful mid-infrared measurement has been achieved.[8]

VCD measurement can also be carried out using a Fourier transform infrared (FT-IR) spectrometer.[9-11] Measurements to date have employed a double-modulation technique in which the high-frequency polarization modulation is superimposed on the Fourier intensity modulations. The same electro-optic components are employed for FT-IR-VCD that were described above for dispersive VCD. Spectra are obtained by first sending the FT-IR-VCD detector signal to a lock-in amplifier tuned to the polarization modulation frequency. The output of the lock-in contains the interferogram of the VCD spectrum which is then Fourier transformed to the yield the final result. FT-IR-VCD measurement is superior to dispersive VCD measurement for the same reasons that FT-IR absorption is superior to conventional IR absorption instruments.[9,12] These advantages have resulted in VCD spectra that are unrivaled in their combination of signal-to-noise (SNR) ratio and spectral resolution. As an example, the mid-infrared VCD and absorbance spectrum of $(-)$-α-pinene measured with a Nicolet 7199 spectrometer system is given in Figure 1.

III. VCD IN AMINO ACID MOLECULES

The importance of amino acids, derivatives of amino acids, and small peptides as regulators and mediators of biological activity is gaining widespread attention in biochemistry in general and in such specialized fields as neurobiology and pharmacology. Of critical interest are the conformations that these molecules adopt in solution as they carry out their specific biological functions. In our laboratory we have undertaken the study of the amino acids, amino acid derivatives, small peptides, and transition metal complexes of these molecules using VCD spectroscopy. The principal aim of these studies is to uncover new stereochemical information concerning the solution structures of these molecules.

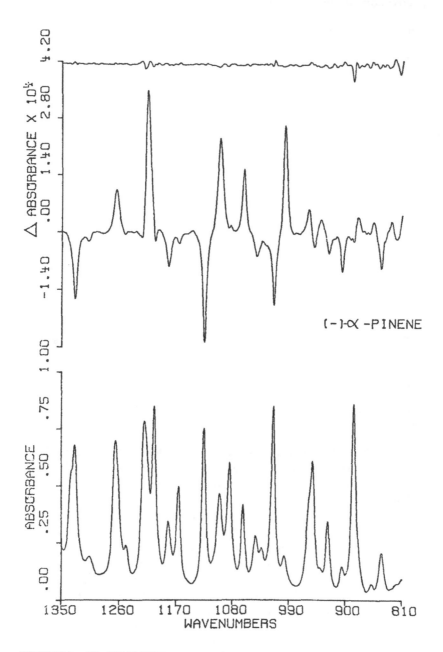

FIGURE 1.   The FT-IR-VCD spectrum of (−)-α-pinene (middle) with the absorbance spectrum (bottom) and estimate of the noise in the VCD spectrum (top). The spectra were obtained using a neat sample using a pathlength of 77 μm.

A. Amino Acids

Most of our VCD work on amino acids has been carried out in the carbon-hydrogen (C-H) stretching region in D_2O solution. In our most recent publication in a series we reported the discovery of a chirality rule for the $C^*_\alpha H$ stretching mode in amino acids and closely related molecules.[13] The rule is based on the observation that the VCD in the C-H stretching region for virtually all of the naturally occurring L-amino acids is biased to positive VCD intensity. The origin of this bias can be traced to the motion of electronic charge that is distinct from the nuclear motion during the methine $C^*_\alpha H$ stretching vibration. Based on evidence that includes the VCD spectra of amino acids

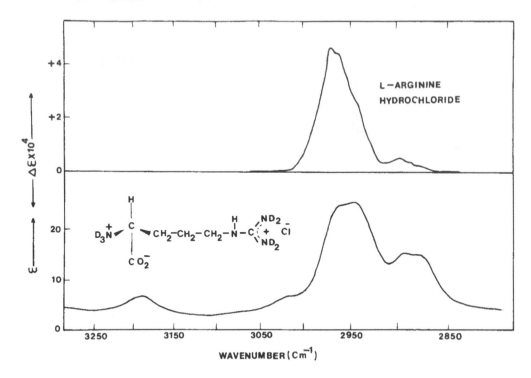

FIGURE 2.   Dispersive VCD spectrum (top) and absorbance (bottom) of L-arginine hydrochloride in D_2O solution. The concentration was 1.4 *M* and the pathlength was 100 μm. The VCD noise uncertainty is $\pm 0.1 \times 10^{-4} \Delta\varepsilon$ units.

as a function of pH (pD) and transition metal complexes of amino acids, we have postulated that the origin of the VCD bias can be further traced to vibrationally induced electronic current in a ring formed by an intramolecular hydrogen bond between the CO_2^- and ND_3^+ groups of the amino acid. Whether or not a ring current is present, there is strong circumstantial evidence to indicate that VCD is particularly sensitive to the presence of intramolecular hydrogen bonds in the amino acids and related molecules. Information of this kind restricts the range of possible solution conformations of these molecules and perhaps more interestingly those of the simple peptides. As an example of the strong VCD bias that can arise in the CH stretching region of amino acids, we present in Figure 2 the dispersive VCD spectrum of L-arginine hydrochloride.[14] Through the bias, the VCD spectrum supports the idea that the hydrochloride is associated with the amine groups on the side chain and not the alpha amino group, ND_3^+, which presumably is associated with the CO_2^- group.

B. Amino Acid Derivatives

In addition to the amino acids and simple peptides, we have carried out VCD studies on a variety of amino acid derivatives. Of particular interest to us are the urethane derivatives of alanine and proline. We have carried out mid-infrared FT-IR-VCD studies in which the carbonyl stretching mode of the urethane group interacts with the carbonyl group of the carboxylic acid moiety. For N-CBZ-L-alanine in $CDCl_3$ solution, both of these carbonyl groups have stretching frequencies in the 1700 to 1750 cm^{-1} range, and the VCD in this region exhibits a sigmoidal shape with positive intensity on the high-frequency side and negative on the low-frequency side of the unresolved carbonyl absorption band. The VCD spectrum is sensitive to the stereochemical orientation of the two carbonyl groups which are held fixed by an intramolecular hydrogen bond. Quantitative information can be obtained using the coupled oscillator model of VCD intensity to interpret the spectrum.[15] Knowledge of the orientation angles of the carbonyl groups provides detailed data that relates to the overall solution phase conformation of this molecule.

C. Transition Metal Amino Acid Complexes

We have obtained VCD spectra in the C-H stretching region for a number of bis Cu(II) amino acid complexes as well as tris Co(III) amino acid complexes.[14] In this region we observe a VCD bias that is always larger than the isolated amino acid spectrum, indicating that an enhanced level of electronic current is present during the methine $C_\alpha^* H$ stretching mode due to the transition metal. The transition metal ion serves a bridge between the amino and carboxylic functional groups and is more efficient in generating a magnetic moment than an intramolecular hydrogen bond. Again, we find an instance in which VCD is very sensitive to the degree of intramolecular interaction between functional groups in a molecule. While other techniques may be sensitive to these interactions, VCD appears to be selectively sensitive to intramolecular interactions in which an electronic bridge is established between groups that are part of the same molecule and hence may be vibration coupled through the rest of the covalent framework.

We have also studied transition metal complexes in the NH stretching region in acidic D_2O solutions.[14] The low pD conditions prevent the exchange of the nitrogen hydrogens in the tris Co(III) complexes of alanine. This molecule exhibits three forms of chirality: (1) configurational by virtue of the sense of the propeller form by the three bidentate amino acid ligands; (2) vicinal by the absolute configuration of the constituent amino acids and; (3) conformational through the stereo-geometry of the amino acid backbone and side chains. In the N-H stretching region of the amino groups of alanine ligated to Co(III), the configurational chirality is the dominant effect whereas in the C-H stretching region, the vicinal effect appears to be the most important.

IV. VCD IN POLYPEPTIDES

Another major area of interest in our research using VCD spectroscopy is polypeptide structure. In Figure 3 we present the FT-IR-VCD spectrum of poly(γ-benzyl-L-glutamate) in $CHCl_3$ solution.[16] The spectral resolution is 4 cm^{-1}, which is higher than previous studies.[17,18] This polypeptide is known to assume the α-helical conformation under these conditions. The amide I band at 1652 cm^{-1} is characteristic of this secondary structure. Also shown in this figure is the C=O stretching band of the benzyl group at 1730 cm^{-1} which exhibits no VCD, and the amide II band at 1550 cm^{-1} which possesses weak negative VCD. The large VCD corresponding to the amide I band shows two major bands and a minor negative band aligned with the low-frequency side of the absorption band. These VCD bands are quite narrow and represent only the residual of broader component bands of opposite VCD intensity that cancel a great deal of their intrinsic intensity. The VCD features can be interpreted on the basis of theoretical descriptions of VCD involving vibrational exciton coupling between the subunits of the polypeptide.[19,20] From considerations based on these theoretical descriptions, the two large VCD features originate from a weak absorption band component on the high-frequency side of the amide I band corresponding to vibrational modes that are polarized perpendicular to the helical axis. The weak VCD feature corresponds to the main absorption maximum and originates in a vibrational mode that is polarized parallel to the helix axis. Presently, we are interested in determining the degree to which the observed VCD spectrum deviates from the theoretical predictions for a perfectly straight infinite helix.

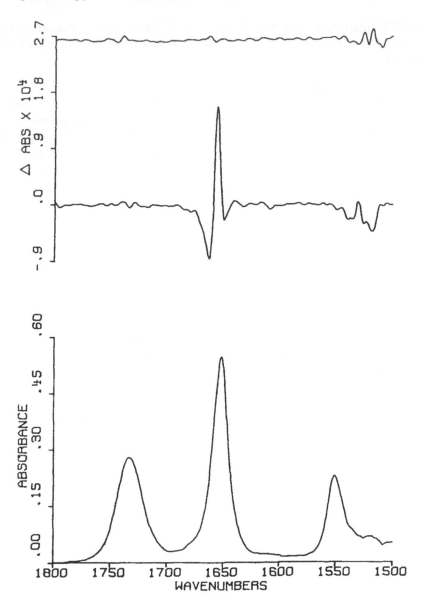

FIGURE 3.   FT-IR-VCD spectrum (middle) of poly(γ-benzyl-L-glutamate) with the absorbance spectrum (bottom) and noise estimate (top). The sampling conditions were resolution, 4 cm^{-1}; concentration in CHCl$_3$, 3 mg/ml; pathlength, 500 μm.

V. FUTURE APPLICATIONS AND DEVELOPMENTS

Vibrational CD is a developing area of molecular spectroscopy. A recent major advance in its availability to a wider field of potential users was the advent of FT-IR-VCD. By merely augmenting an FT-IR spectrometer with a photoelastic modulator and a lock-in amplifier it is possible to carry out state-of-the-art measurements of VCD. Previously, one needed to invest a year or more of instrument construction to match the capability of dispersive VCD units. As more researchers take up the study of VCD, its range of applications will no doubt grow.

In addition to extending our work with amino acids, peptides, and polypeptides, other areas of biological applications will likely be developed. Sugars[21] and chiral phosphates[22] have been the subjects of VCD studies and nucleic acid molecules and polymers represent an obvious new area of investigation. In our group, small bioactive peptides that act as neurotransmitters and hormonal regulators are long-range targets for investigation. Although these molecules are comprised of just a few amino acid subunits, their conformational freedom is extensive and difficult to describe in solution. NMR and ordinary vibrational spectroscopy are making some headway, but it is likely that VCD spectra will provide unique sets of conformation constraints based on both empirical and theoretical analyses of the VCD spectra.

A number of key areas of instrumental capability need to be developed before VCD can gain wider access to biological molecules. The first is instrument performance in terms of signal quality, spectral range, and the speed of spectral acquisition. These performance characteristics are governed primarily by source intensity, optical throughput, photoelastic modulator efficiency, and detector response quality. Improvements in these areas are appearing at a steady rate and there is every reason to expect improvements in the future. For instance the use of a liquid helium cooled copper-doped germanium detector can be expected to provide longer wavelength coverage and improved SNR characteristics compared to HgCdTe. Another area of improvement which is related to instrument performance is the ability to obtain spectra from small amounts of sample, to observe VCD spectra in the presence of strong solvent absorption such as encountered in D_2O and H_2O, and to be able to obtain quality spectra from a single enantiomer. Recently, we achieved positive results in the last of these points using a method of varying the concentration at fixed pathlength.[11] Finally, recent software developments that lead to enhancements in the spectral resolution of overlapping bands will likely have a significant impact on VCD applications. By itself, VCD often resolves overlapping spectral components, e.g., Figure 3. We have recently demonstrated that Fourier selfdeconvolution can be successfully applied to VCD spectra to further resolve overlapping bands.[23] Since adjacent bands of opposite intensity cancel when they overlap, the effects of deconvolution can be quite dramatic in VCD compared to absorption spectra in certain instances.

As a final point we note that Raman optical activity may serve as an alternative source of stereo-sensitive spectral information in the coming years. ROA at the moment is not as readily applicable to biological problems due to technical obstacles in obtaining quality spectra. However, ROA has a number of intrinsic advantages such as broad vibrational spectra coverage, relative ease in aqueous sampling, and structural selectivity afforded by the resonance Raman effect. Thus, future applications of VCD spectroscopy to biological systems will likely be complemented and supplemented by parallel measurements using ROA spectroscopy.

ACKNOWLEDGMENT

Acknowledgment is given to the National Institutes of Health (GM-23567) and the National Science Foundation (CHE-83-02416) for supporting the research on vibrational optical activity described above from our laboratory.

REFERENCES

1. Nafie, L. A., Experimental and theoretical advances in vibrational optical activity, in *Advances in Infrared and Raman Spectroscopy*, Vol. II, Clark, R. J. M. and Hester, R. E., Eds., John Wiley & Sons, New York, 1984, 49.

2. Barron, L. D., *Molecular Light Scattering and Optical Activity*, Cambridge University Press, London, 1983.
3. Mason, S. F., *Molecular Optical Activity and the Chiral Discriminations*, Cambridge University Press, London, 1983.
4. Mason, S. F., Molecular handedness and the origins of chiral discriminations, *Int. Rev. Phys. Chem.*, 3, 217, 1983.
5. Bayley, P., Circular dichroism and optical rotation, in *An Introduction to Spectroscopy for Biochemists*, Brown, S. B., Ed., Academic Press, New York, 1980, 148.
6. Moscowitz, A., Theoretical aspects of optical activity. I. Small molecules, *Adv. Chem. Phys.*, 4, 67, 1962.
7. Nafie, L. A., Infrared and Raman vibrational optical activity, in *Vibrational Spectra and Structure*, Vol. 10, Durig, J. R., Ed., Elsevier, Amsterdam, 1981, 153.
8. Keiderling, T. A., Vibrational circular dichroism, *Appl. Spectrosc. Rev.*, 17, 189, 1981.
9. Nafie, L. A. and Vidrine, D. W., Double modulation Fourier transform spectroscopy, in *Fourier Transform Infrared Spectroscopy*, Vol. 3, Ferraro, J. R. and Basile, L. J., Eds., Academic Press, New York, 1982, 83.
10. Lipp, E. D., Zimba, C. G., and Nafie, L. A., Vibrational circular dichroism in the mid-infrared using Fourier transform spectroscopy, *Chem. Phys. Lett.*, 90, 1, 1982.
11. Lipp, E. D. and Nafie, L. A., Fourier transform infrared vibrational circular dichroism: improvements in methodology and mid-infrared spectral results, *Appl. Spectrosc.*, 38, 20, 1984.
12. Griffiths, P. R., Fourier transform infrared spectrometry: theory and instrumentation, in *Transform Techniques in Chemistry*, Griffiths, P. R., Ed., Plenum Press, New York, 1978, 109.
13. Nafie, L. A., Oboodi, M. R., and Freedman, T. B., Vibrational circular dichroism in amino acids and peptides. VIII. A chirality rule for methine C^*-H stretching modes, *J. Am. Chem. Soc.*, 105, 7449, 1983.
14. Oboodi, M. R., *Vibrational Circular Dichroism of Amino Acids and Some of Their Transition Metal Complexes*, Ph.D. dissertation, Syracuse University, Syracuse, N. Y., 1982.
15. Holzwarth, G. and Chabay, I., Optical activity of vibrational transitions: a coupled oscillator model, *J. Chem. Phys.*, 57, 1632, 1972.
16. Lipp, E. D. and Nafie, L. A., unpublished data, 1983.
17. Singh, R. D. and Keiderling, T. A., Vibrational circular dichroism of poly (-benzyl-L-glutamate), *Biopolymers*, 20, 237, 1981.
18. Lal, B. B. and Nafie, L. A., Vibrational circular dichroism in amino acids and peptides. VII. Amide stretching vibrations in polypeptides, *Biopolymers*, 21, 2161, 1982.
19. Snir, J., Frankel, R. A., and Schellman, J. A., Optical activity of polypeptides in the infrared. Predicted CD of the amide I and amide II bands, *Biopolymers*, 14, 173, 1975.
20. Schellman, J. A., Optical activity of polypeptides in the infrared: symmetry considerations, in *Peptides, Polypeptides and Proteins*, Blout, E. R., Bovey, F. A., Goodman, M., and Lotan, N., Eds., John Wiley & Sons, New York, 1974, 320.
21. Marcott, C., Havel, H. A., Overend, J., and Moscowitz, A., Vibrational circular dichroism and individual chiral centers. An example from the sugars, *J. Am. Chem. Soc.*, 100, 7088, 1978.
22. Su, C. N., Keiderling, T. A., Misiura, K., and Stec, W.J., Midinfrared vibrational circular dichroism studies of cyclophosphamide and its cogeners, *J. Am. Chem. Soc.*, 104, 7343, 1982.
23. Lipp, E. D. and Nafie, L. A., Application of Fourier self-deconvolution to vibrational circular dichroism spectra, *Appl. Spectrosc.*, 38, 774, 1985.

Chapter 7

BIOMEDICAL APPLICATIONS OF RAMAN SCATTERING

Richard Mendelsohn

TABLE OF CONTENTS

I. INTRODUCTION

Unlike some of the other techniques described elsewhere in this volume, Raman spectroscopy is a flourishing, well-developed subject. Its theoretical basis and applications to biochemistry and biophysics have been reviewed elegantly and often.[1-15] This fact, coupled with a paucity of what might be termed truly biomedical applications, presents the current author with a dilemma. Another review of biochemical applications, especially in light of the excellent recent volume by Carey,[10] seems superfluous, while a review of the sparse biomedical applications is not yet indicated. As a compromise, the following brief chapter is directed toward three ends:

1.   To present the reader with a brief theoretical overview (at a simple level) of Raman and resonance Raman scattering, especially with regard to the type of useful structural information available from each approach and the differences between them.
2.   To describe several of the new experimental approaches, and emphasize their potential utility in biomedical applications.
3.   To illustrate, via reference to a specific application (biomembrane structure), the various tactics that have been employed to extract information of potential interest to biomedical researchers, who will perhaps be thus motivated to look more deeply into this technology.

As implied above, no attempt at complete, comprehensive coverage will be made. Further details can be extracted from the references.

II. THEORETICAL BASIS OF RAMAN SCATTERING

A. The Ordinary Raman Effect

The Raman effect arises because light shining on a sample (and not absorbed by it) exerts a force on the electrons and displaces them from their equilibrium positions around the atomic nuclei. A dipole moment is thereby induced in the molecule. The magnitude of this induced moment is proportional to the electric field strength; thus

$$\vec{M} = \alpha \vec{E} \tag{1}$$

The proportionality constant connecting the electric field vector $\vec{E}$ with the induced dipole moment vector $\vec{M}$ is a 3×3 tensor quantity known as the polarizability, α. It is simplest to consider three scalar equations generated by expanding Equation 1 as follows:

$$M_x = \alpha_{xx}E_x + \alpha_{xy}E_y + \alpha_{xz}E_z$$
$$M_y = \alpha_{yx}E_x + \alpha_{yy}E_y + \alpha_{yz}E_z$$
$$M_z = \alpha_{zx}E_x + \alpha_{zy}E_y + \alpha_{zz}E_z \tag{2}$$

The quantities $\vec{M}$, α, and $\vec{E}$ vary with time. Thus, for the electric field vector of the incident light (assuming light polarized in the xy plane and propagating in the Z-direction)

$$E_z(t) = E_z^o \cos 2\pi\nu_L t \tag{3}$$

where E_z^o is the maximum of $E_z(t)$, and ν_L = the frequency of the incident laser light. For simplicity (rarely achieved for low-symmetry macromolecules), it is assumed that only one component of the tensor (say α_{zz}) is nonzero for the molecule under consideration, so that the time dependence of M is expressed as

$$M_z(t) = \alpha_{zz}E_z^o \cos 2\pi\nu_L t \tag{4}$$

However, α_{zz} itself is a function of time. Thus, for small changes in molecular geometry (such as occur when a molecule vibrates),

$$\alpha_{zz}(t) = \alpha_{zz}^o + \frac{d\alpha_{zz}}{dq} \Delta q \cos 2\pi\nu_{vib}t \tag{5}$$

α_{zz}^o is the static polarizability arising from the equilibrium geometry of the molecule, ν_{vib} is the molecular vibrational frequency, and Δq is the maximal change in internuclear configuration which causes the change in polarizability during the vibration of interest. This implies that

$$\Delta q(t) = \Delta q \cos 2\pi\nu_{vib}t$$

In practice, one must sum over all $3N - 6$ ($3N - 5$ for a linear molecule) normal modes of vibration and over all components of α. For the simple case at hand, the expression for M(t) becomes

$$M_z(t) = \alpha_{zz}^o E_z^o \cos 2\pi\nu_L t + \frac{d\alpha_{zz}}{dq} \Delta q E_z^o \cos(2\pi\nu_{vib}t)\cos(2\pi\nu_L t) \tag{6}$$

On rearrangement and making use of the trigonometric identity

$$\cos A \cos B = 1/2 (\cos (A + B) + \cos (A - B)$$

the desired result is achieved:

$$M_z(t) = \alpha_{zz}^o E_{max} \cos 2\pi\nu_L t + 1/2 \frac{d\alpha_{zz}}{dq} \Delta q E_z^o$$

$$\{\cos[2\pi(\nu_L + \nu_{vib})t] + \cos[2\pi(\nu_L + \nu_{vib})t]\} \tag{7}$$

What are the consequences of this equation?

According to classical electromagnetic theory an oscillating dipole $M_z(t)$ will radiate light at the frequency of oscillation. Thus, incident light impinging on a molecule will be scattered with a frequency dependence (for each vibrational mode) consisting of three components:

1. A component scattered at the incident light frequency, which depends for its magnitude the static polarizability α_{zz}^o. This is Rayleigh scattered light, a nuisance in the current experiment, but widely used in macromolecular size distribution determination.

2. A component scattered at a frequency $\nu_l - \nu_{vib}$; that is, shifted to lower frequency from the incident light by an amount equal to a vibrational quantum of the scattering molecule.

3. A component scattered at a frequency $\nu_l + \nu_{vib}$; that is, shifted to higher frequency from the incident light by an amount equal to a vibrational quantum. Processes

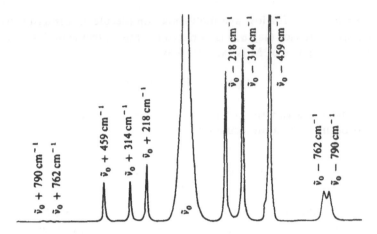

FIGURE 1.   Rayleigh and Raman spectra of carbon tetrachloride in the liquid state excited by an Argon ion laser with a wavelength of 4880 A. The peak marked ν_0 arises from Rayleigh scattering of the excitation frequency. The lines frequency shifted by 218 cm^{-1}, 314 cm^{-1}, and 459 cm^{-1} to the high- and low-frequency side of the Rayleigh scattered line represent anti-Stokes and Stokes Raman lines from fundamental vibrational motions of the molecule. The doublet at 762, 790 cm^{-1} is a Fermi resonance phenomenon and involves only one fundamental vibration. (From Long, D. A., *Raman Spectroscopy*, McGraw-Hill, New York, 1977, 5. With permission.)

2 and 3 represent the Stokes and anti-Stokes Raman scattered light, respectively. The Raman spectrum of a molecule is the entire pattern of frequency-shifted light. As can be seen from Equation 7, at least one component of the polarizability tensor must change during a particular normal molecular vibration in order for that motion to be Raman-active.

Thus, in the Raman spectrum one observes a strong feature at the excitation frequency (corresponding to the Rayleigh line), as well as lines shifted to both higher and lower frequencies by an energy equal to a vibrational quantum of the scattering molecule. One such pair of lines is observed for every motion that is Raman-active. The situation is depicted in Figure 1.

What is the physical picture of polarizability? It is rather simply, the measure of the deformability of a molecule in an electric field. In order for a chemical bond to undergo a large change in polarizability during its vibration, there must be regions of relatively high electron density along the line between the atoms in the bond. Thus, bonds between atoms of similar electronegativities will usually show large polarizability changes during a stretching motion of the bond. Atomic groupings such as C=C, C—C, S—S, N—N, C—S, etc. are expected to be strong Raman scatterers. Readers may note that homopolar bonds are weak absorbers of infrared radiation. (If the molecular system is sufficiently symmetric, absorption in some cases may be rigorously forbidden.) Thus, certain modes which scatter strongly in the Raman spectrum may be weak in the IR. Similarly, bonds which are polar and therefore have a rather asymmetric distribution of charge, such as C=O bonds, are strong absorbers in the IR but rather weak Raman scatterers. The two techniques thus yield complementary information.

The classical expression for the Raman effect (Equation 7 above), incorrectly predicts that the Stokes and anti-Stokes Raman lines have equal intensities. A "semi-quantum mechanical" description of events superimposes the quantization of molecular vibrations on the above classical picture. Thus, the allowed vibrational wavefunctions (in the harmonic oscillator model for vibrating systems) are given by

$$\Psi_v = N_v \exp(-1/2\alpha q^2) \, H_v(\sqrt{\alpha q})$$

where $\alpha = 2\pi \sqrt{\mu k/h}$, q = the normal vibrational coordinate, H_v is the Hermite polynomial of degree v, N_v is the normalizing factor, k is the force constant for harmonic motion, and μ is the reduced mass.

For typical vibrational energies (600 to 2000 cm^{-1}), most molecules are in their ground (v = 0) vibrational states. Stokes Raman lines, in this description, arise from those molecules which are excited to the v = 1 level during the scattering process. The anti-Stokes Raman lines arise from those (fewer) molecules initially in the v = 1 state which after the scattering event, end up in v = 0. The relative intensity of the Stokes and anti-Stokes lines is approximately correctly predicted to be dependent on the relative numbers of molecules initially in the v = 0 and v = 1 levels; i.e., a Boltzmann factor. This ratio is

$$N_1 = N_0 \exp(-h\nu_{vib}/kT)$$

where N_1 = number of molecules in the v = 1 level, N_0 = number of molecules in the v = 0 level, and h, T, and k are Planck's constant, the absolute temperature, and the Boltzmann constant, respectively. It is common practice to measure only the Stokes Raman spectrum, as the anti-Stokes spectrum is substantially weaker and contains no new information.

B. The Resonance Raman Effect

In the above description of Raman scattering, it was assumed that the incident light energy lies well below that required for electronic excitation of the molecule. The development of resonance Raman scattering as an important biophysical tool evolved under conditions precisely to the contrary; so that in a resonance Raman experiment the excitation frequency lies within the bandwidth of an electronic molecular absorption band. It is noted that there is a continuous transition between ordinary Raman scattering and resonance Raman scattering. For example, when the excitation occurs on the low-energy tail of the absorption band, the experiment is said to be in the preresonance Raman regime. The situation is depicted in Figure 2.

Under conditions of resonance Raman scattering, the expansion of the polarizability tensor into a Taylor series as in Equation 5 is invalid, and the expression for the intensity of scattered light for a transition from state m to state n is given by

$$I = A \, (\nu_o + \nu_{mn})^4 \, I_o \sum_{\sigma,\rho} |(\alpha_{\sigma\rho})_{mn}|^2$$

where $\alpha_{o\rho}$ is one Cartesian component of the polarizability tensor connecting states m and n, ν_{mn} is the frequency of the vibrational transition, ν_o is the frequency of the incident light, A is a constant, and I_o is the incident light intensity. The formal quantum mechanical expression for $(\alpha_{o\rho})_{mn}$ is

$$(\alpha_{\sigma\rho})_{mn} = \frac{1}{h} \sum_r \frac{<n|\mu_\rho|r><r|\mu_\sigma|m>}{\nu_{rm} - \nu_o + i\Gamma_r} + \text{other terms} \tag{8}$$

where the sum r is taken over all molecular vibronic states, ν_o is the excitation frequency, h is Planck's constant, and ν_{rm} is the frequency difference between the state r and the ground state m. The integrals in the numerator are the dipole moment matrix elements connecting the states indicated via the dipole moment operator along the indicated direction, either μ_ρ or μ_σ, Γ_r is a constant related to the line widths (lifetimes) of the vibronic state r.

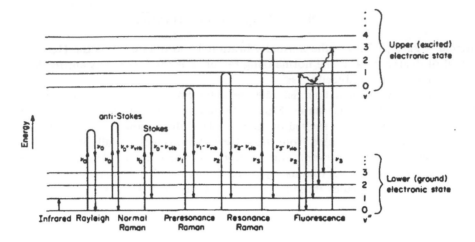

FIGURE 2. A schematic depiction of the Raman scattering process showing some of the possible results of a photon-molecule interaction. The upper set of vibrational energy levels is that of the excited electronic state v′, while the lower set belongs to the ground electronic state. Raman and resonance Raman scattering are two-photon processes which occur so rapidly as to be considered concerted single events. Fluorescent emission, also a two-photon process, is substantially slower and is depicted as two one-photon events occurring sequentially. (From Carey, P. R., *Biochemical Applications of Raman and Resonance Raman Spectroscopies*, Academic Press, New York, 1982, 22. With permission.)

Equation 8, in its most general form, has an infinite number of terms which contribute to the Raman scattering intensity. In fact, there is one for each vibronic state r. Various schemes for simplifying this equation have been summarized,[10] and will not be repeated here. Several points are relevant to gaining a qualitative understanding of Equation 8:

1.   As the laser frequency approaches the frequency of the vibronic transition, the denominator for a particular term α_{op} becomes relatively small, so that the intensity of scattered light as given by Equation 8 increases. In addition, a small number of terms in the infinite sum will dominate the polarizability expression. Each of these corresponds to a different electronic or vibronic transition.

2.   Those vibrations which are enhanced in intensity are those which have substantial geometry change along a direction or region of the molecule where the electronic transition is located. This is a crucial consideration, the physics of which has been discussed in many studies.[16-18] For example, in the lowest-energy electronic transition of β-carotene, the excitation involved is a π-π* promotion parallel to the long axis of the molecule. Only those vibrations having substantial contributions from nuclear motions along coordinates in this direction are enhanced in Raman intensity under resonance conditions. Thus, the resonance Raman spectrum of β-carotene is dominated by C=C and C—C stretching vibrations, whereas the C—H stretching modes are absent.

3.   It is clear from the above, that chromophore-containing systems (or at least, molecular groupings the absorption bands of which correspond with available laser lines) are required for resonance Raman studies. Thus, there must be an absorbing species present in the native biological system. In the resonance Raman spectrum of rhodopsin, the only vibrations observed are those of the retinylidenelysine moiety. The vibrations of the opsin (not resonance enhanced) are buried in the background noise.

If the system has no chromophore, one must be introduced. This approach has been pioneered by Carey[10,15] in studies of enzyme mechanisms by resonance Raman spectroscopy.

In summary, the differences between Raman and resonance Raman scattering are as follows. In Raman scattering, all normal modes of a molecule may contribute to the spectrum, the relative intensities depending solely upon $d\alpha_{xx}/dq$ in Equation 5. Vibrations of groups with high electron density between nuclei are expected to be the strongest. In resonance Raman scattering, certain chromophoric vibrations may be selectively enhanced when the laser excitation lies within the envelope of a molecular electronic absorption band.

C. Interpretation of Biomolecular Raman Spectra

The level of spectral data interpretation for biomolecules is generally much less sophisticated than that applied to small molecules in the gas phase. An empirical procedure has evolved in which spectral interpretation is based on extrapolation from a large body of data from small relevant model systems. Data sets from the latter are collected in as complete a fashion as possible, and include both infrared and Raman spectral frequencies, intensities, polarizations (see References 8, 10, and 16 for a discussion of polarization properties of Raman scattered light), and isotopic substitutions, if possible. An attempt is then made to assign the spectral features to particular molecular vibrations based on the known structure of the model system. The normal modes of the model system can then be calculated. At this point, the nature of each molecular vibration, including the amplitudes and displacements of every atom during the vibration, is known.

When the previous methods have been employed for many structurally related molecules, it has been determined that certain vibrations in large molecules only involve the motion of a few atoms. These atoms are said to possess a characteristic vibrational frequency called a group frequency. Small variations in such a mode can often be correlated with primary or secondary structural features of interest. The application of the group frequency concept to organic chemistry has been discussed in several excellent volumes.[19,20] These group frequencies may sometimes be transferred from the model to the biopolymer system, for determination of functional groups present in the latter. It may also be possible to use small shifts in the frequencies to obtain information about biopolymer secondary structure. The reliability of the conclusions drawn depends upon the transferability of the spectra-structure correlations from simple to complex systems. The level of confidence in the entire procedure is enhanced as more and more model systems are investigated.

The above procedure has been successful in determination of spectra-structure correlations from ordinary Raman spectra, as will be illustrated below. The procedure tends to break down in resonance Raman studies of complex π-electron systems such as the iron-protoporphyrin IX chromophore of heme proteins. In such cases, many internal coordinates may contribute to each vibration so that accurate spectra-structure correlations require detailed normal coordinate calculations.[21]

To illustrate the utility of Raman spectroscopy in biomedically related areas, the application of the technique to the structure of biological membranes will be outlined. Leading references to other biochemical applications will be given.

III. NEW TECHNIQUES IN RAMAN SPECTROSCOPY

The application of Raman spectroscopy to biomedical problems has been hindered by two drawbacks of the technique. First, the intensity of scattered light in the ordinary Raman effect is weak. Compared with fluorescent emission, Raman scattering is about

10^6 times weaker on a molar basis. The ability to collect data in the presence of small amounts of fluorescent constituents is therefore limited. The second drawback is that native biological samples (cells, tissues, etc.) are likely to be spatially nonuniform. Thus, any ordinary Raman signal obtained is an average over a number of different physical species.

The resonance Raman effect provides a means of overcoming sensitivity difficulties due to low intensity of scattered light. However, this solution to the intensity problem is not universal. As discussed above, the molecule of interest must be chromophoric; i.e., must absorb the incident radiation. At first glance, it might be thought that clever selection of the incident radiation would enable any molecule to act as a chromophore. However, as the excitation frequencies are increased toward the UV spectral regions, the chance of sample destruction (especially of native tissue specimens) becomes significant. Nevertheless, some results have been reported with UV excitation for nucleic acids and proteins,[22-24] and further work in this area may prove fruitful biomedically.

In addition to resonance Raman scattering, three new types of Raman experiments, which may serve to alleviate some of the difficulties inherent in biomedical samples, have been developed in recent years. Each is discussed in the following sections.

A. Coherent Anti-Stokes Raman Scattering

A significant drawback to the acquisition of acceptable Raman spectra from native biological samples is the inevitable occurrence of small amounts of fluorescent species, which tend to obscure the Raman spectrum. The phenomenon of coherent anti-Stokes Raman scattering (CARS) may provide a way around this difficulty. The effect is observed when two approximately colinear laser beams of frequencey w_1 and w_2 ($w_1 > w_2$) are directed on a sample that has a Raman-active vibration at $w_1 - w_2$. Intense coherent scattering is observed at $2w_1 - w_2$. Since the scattered light is on the anti-Stokes side of the exciting line and is coherent, possible interference from fluorescence, which is incoherent and on the Stokes side of the excitation frequency, is minimized. Due to the coherence of the scattered light, the detector may be placed far enough from the sample so that fluorescent emission, which is incoherent, is spatially discriminated against. The phenomenon was initially observed by Maker and Terhune,[25] and has since been reviewed by Hudson.[26] Although intense laser light is needed to observe the phenomenon (leading to the lively possibility of sample deterioration in the laser beam), the feasibility of using the CARS approach for acquiring protein spectra was shown several years ago by Spiro and co-workers.[27]

A recent application is the acquisition of a CARS spectrum from the (highly fluorescent) flavin chromophore in flavodoxins from *Desulfovibrio gigas* and *D. vulgaris* as well as from the 6,7-dimethyl-8-ribityllumazine chromophore from the bioluminescent protein of *Photobacterium phosphoreum*.[28] An additional application[29] uses a variant of CARS under resonance conditions to examine the vibrational structure of excited electronic states in pyrene.

B. The Raman Microprobe

An obvious difficulty with biological samples is microscopic heterogeneity. The normal spot size of a focused laser beam is about 0.1 mm, which would yield Raman spectra averaged over a diverse variety of structural elements. The coupling of a microscope and an image intensifier with a Raman spectrometer permits the examination of biological materials *in situ* and provides a solution to this problem. Microscope objectives produce a 1-μm spot on a sample with a depth of focus of 5 μm. Spatial resolution to this level may therefore be achieved, assuming proper design of the Raman optical system. To prevent sample degradation during the long exposure time when spectra are recorded using a monochromator and single photomultiplier, the latter is replaced by a multichannel detector. This approach, pioneered by Delhaye and co-workers in France,[30,31] can yield a two-dimensional "image" of an object according to the Raman spectrum of its constituents. The potential of a nondestructive method, which combines chemical identification of a structure with physical location in a complex environment is clear. Although biomedical applications of the method are scarce, Buiteveld et al.[32] have examined microscopic particles in human lung tissue. Paraffinized unstained sections were deposited on microscope slides and 5145 Å radiation from an Argon ion laser was focused to diameters of laser spots ranging from 1.6 to 4 μm. Particles of $CaCO_3$ (calcite) were unambiguously identified in the tissue.

C. Surface-Enhanced Raman Scattering

A further mechanism for intensity enhancement of Raman signals offers some hope for biomedical problems. It was initially noted that pyridine adsorbed on a silver surface that had been electrochemically roughened produced a Raman signal many orders of magnitude larger than expected for the concentration of molecules. The phenomenon has been observed since by many groups for a variety of species adsorbed usually on silver but occasionally on copper surfaces. The phenomenon has been termed surface-enhanced Raman scattering (SERS). Although complete agreement as to the physical mechanism responsible for the enhancement of signals has not been achieved, several factors are implicated. Burstein and Chen[33] have reviewed the role of submicroscopic (<100 Å) surface roughness in the intensity enhancement process. Wood and Klein[34] have presented evidence that the enhancement that occurs when CO is adsorbed on a rough Ag surface is reduced by a factor of 500 when the surface is annealed. In addition, Creighton et al.[35] studied the SERS phenomenon in molecules adsorbed to Ag or Au sols. They noted that the enhancement was maximal at the wavelength of the resonance peak in the Mie scattering by the sol particles, and suggested that the enhanced scattering is due to excitation of surface plasmons or structures on the surface. Moskovitz[36] further indicated that optical excitations of the transverse collective electron resonances could explain the wavelength dependence of SERS. A good summary of this entire complex topic has been given by Burstein and Chen.[33]

Several problems must be overcome before SERS can be applied as a method for enhancement of all Raman signals. First, the spectrum of the adsorbed species is substantially altered from the free substance. Van-Duyne[37] has concluded that only vibrations directed perpendicular to the metal surface are enhanced. These factors will lead to problems both in the identification of species and in the development of spectra-structure correlations. In addition, the types of surfaces upon which this enhancement occurs must be extended in order to ensure the greatest possible flexibility in biomedical sample preparation. In spite of the above problems, spectra of nucleic acids adsorbed on Ag surfaces have been reported at concentrations of 10^{-4} M.[38]

IV. APPLICATIONS TO BIOLOGICAL MEMBRANE STRUCTURE

As an illustration of the approach used to extract structural information from Raman spectra of biomolecules, applications of the technique to the study of biological membranes will be described. The current general model for biomembrane structure can be found in any current biochemistry text (see, for example, Reference 39). The main chemical constituents of biomembranes are lipids and proteins, so that spectra-structure correlations for these two molecular classes will be reviewed first.

A. Protein Structure

Understanding the Raman spectra of proteins requires detailed knowledge of the spectra of (1) the 20 or so amino acids commonly found in proteins and (2) synthetic

Table 1
VIBRATIONAL FREQUENCIES OF THE PEPTIDE BOND
IN N-METHYLACETAMIDE

Raman	Infrared	Assignment
3300	3300	N−H stretching
	3110	
1657	1653	Amide I (80% C=O stretch)
	1567	Amide II (60% N−H in-plane bend, 40% C−N stretch)
1298	1299	Amide III (40% C−N stretch, 30% N−H in-plane bend, 20% CH₃−C stretch)
	725	Amide V (N−H out-of-plane bend)
628	627	Amide IV (40% O=C−N bend, 30% CH₃−C stretch)
600	600	Amide VI (C=O out-of-plane bend)

Note: Frequencies in cm⁻¹.

homopolypeptides of known secondary structure. This procedure works through relationships — quantitative where possible — between the geometry of these model compounds and the spectroscopic data.

1. Peptide Bond Vibrations

Knowledge of peptide bond vibrations originated with work on the model compound
N-methyl acetamide, as summarized in the normal coordinate calculations of Miyazawa et al.[40] (Table 1). For Raman spectroscopic studies of protein structure, the most
important peptide frequencies are those denoted Amide I (80% C=0 stretch) and
Amide III (40% C−N stretch, 30% N−H in-plane bend, and 20% CH₃−C stretch). The
other modes (Amide II, IV, V, and VI) are less important. Amide II, for example,
although strong in the infrared spectrum of proteins, is weak or absent in the Raman
effect, while Amides IV to VI are often weak and/or overlapped by side-chain vibrations.

To explore the sensitivity of the peptide bond frequencies to changes in secondary
structure, homopolypeptides are useful as model compounds. When a peptide bond is
located in regions of regular and well-defined secondary structure, the peptide bond
frequencies enumerated in Table 1 are altered due to inter- and intrachain coupling
effects through hydrogen bonds as described by Miyazawa and Blout.[41] A localized
vibration placed into a region of regular secondary structure has its frequency modified
by inter- and intrachain interactions as described by

$$\nu(\delta,\delta') = \nu_o + \sum_s D_s\cos(s\delta) + \sum_{s'} D'_s\cos(s'\delta') \qquad (9)$$

where ν_o is the unperturbed frequency; the second and third terms are due to the intra-
and interchain vibrational interactions, respectively; δ is the phase angle between adjacent groups for the motion under consideration; and the interaction coefficients and
the interaction coefficients D_s refer to the s th neighbors in the chain. δ' and D_s' pertain
to interactions through interchain hydrogen bonds and have meanings similar to those
of δ and D_s, respectively. To apply Equation 9, a knowledge of those phase angles δ,
δ', which give rise to Raman-active modes for particular homopolypeptide secondary
structures must be available.

As an example of the procedure, in an α-helix there are 18 peptide bonds per 5 turns
of helix while interchain hydrogen bonds are found between each pair of every third
neighbor. It is verifiable from group theory that the frequencies of the Raman-active

Table 2

VIBRATIONAL FREQUENCIES CHARACTERISTIC OF THE PEPTIDE BOND
IN VARIOUS CONFORMATIONS

Structure	Amide I		Amide II		Amide III	
	Raman	IR	Raman	IR	Raman	IR
−Helix	1655(s)[a]	1655(s)	—	1550(s) 1510(w)	1275 ± 20(m)	1275 ± 20(m)
−Antiparallel pleated sheet	1670(s)	1630(s) 1685(w)	—	1530(s)	1230(s)	1236(m)
Random-coil	1665(m, br)	1660(m, br)	—	1540(w)	1250(m)	1250(m)

Note: Frequencies in cm⁻¹.

[a] s, m, w, and br represent strong, medium, weak, and broad, respectively. Amide II absent in Raman.

modes occur = 0, 100, and 200°, respectively (corresponding in group theoretical language to the A1, E1, and E2 modes of vibration). For homopolypeptides of α-helical structure it is found that these three modes occur with very similar frequencies. This arises because the magnitudes of D and D′ are similar but of opposite sign. The α-helical Amide I frequencies all appear at 1650 ± 5 cm⁻¹. For the antiparallel pleated sheet, a unit cell contains four peptide groups, two each from two adjacent chains. The values of δ and δ' permitted are each 0 or 180°.

From Equation 9, four frequencies for each normal mode are expected, as follows:

$$\nu(o,o) = \nu_o + D_1 + D_1'$$

$$\nu(o,\pi) = \nu_o + D_1 - D_1'$$

$$\nu(\pi,o) = \nu_o - D_1 + D_1'$$

$$\nu(\pi,\pi) = \nu_o - D_1 - D_1' \qquad (10)$$

All four frequencies are expected to be Raman active, although only one — the (o,o) — dominates the spectrum of homopolypeptides and appears in the 1665 to 1670 cm⁻¹ region. In the infrared spectral region they are all allowed, although only two strong bands are normally observed near 1680 and 1630 cm⁻¹. It is clear therefore that the α-helical secondary structure (single Amide I Raman band near 1650 cm⁻¹), is easily differentiated from an antiparallel pleated sheet (single band from 1665 to 1670 cm⁻¹). Similar considerations apply to the IR data.

Although the Amide I frequency in the Raman spectrum is usually free from interference from other vibrations in proteins, it may be overlapped by the broad water band (H−O−H bending) at 1640 cm⁻¹ in aqueous solutions. Replacement of H_2O by D_2O (the corresponding bending vibration of which occurs at 1210 cm⁻¹) will uncover the Amide I′ frequency (C=O stretch of the C−N grouping in peptide bonds) shifted to lower values (from Amide I) by 5 to 10 cm⁻¹ due to the deuterium substitution.

Analysis of the homopolypeptide spectral splitting patterns for various secondary structures has been given by Koenig.[42] The resultant spectra-structure correlations are summarized in Table 2. It is noted that the Amide III frequency is substantially more sensitive than the Amide I as a conformational probe in the Raman effect, as it spans a wider range of frequencies. Several studies, carried out a decade or so ago, demonstrate the general utility of the peptide bond modes as structure probes. For example, Frushour and Koenig[43] have monitored the mechanically induced transformation from

α-helical to β-pleated sheet in poly-L-alanine, while Yu et al.[44] have monitored a similar event in poly-L-lysine. More recent studies have centered around peptide bonds in short peptides, particularly in search of spectra-structure correlations for conformations such as the β-bend.[45]

Difficulties arise in extending the correlations derived from homopolypeptides to problems of protein structure. The group theoretical arguments illustrated earlier used to predict the splitting of each peptide bond vibration into several components in ordered model systems may not be applicable to globular proteins. The short lengths of ordered segments in the latter as well as the possible presence of different side chains will in principle remove all symmetry and destroy the inter- and intrachain coupling. Furthermore, these segments may have a range of hydrogen-bond distances and angles for the peptide groups rather than the precisely fixed values of these parameters in ordered homopolypeptides, so that the latter may not serve as completely reliable models for estimates of various types of secondary structure by Raman spectroscopy.

The thrust of the initial work in the application of the method has been to obtain empirical correlations between protein conformation and peptide bond frequencies. For example, Yu and Liu[46] have examined the polypeptide hormone glucagon in three different conformational states while Lord and co-workers have examined the effect of various types of denaturation processes on the Raman spectrum of both lysozyme and ribonuclease A.[47,48]

Recent applications have centered around the determination of protein secondary structure in situations not easily amenable to other spectroscopic techniques. Dunker et al. have examined complexes of the β-protein of fd phage reconstituted into a model membrane system.[49] In a biomedically related application, Yu et al. and Mizuno et al.[50,51] demonstrated the feasibility of obtaining Raman spectra of an intact lens in a whole eyeball.

Several approaches have been introduced for the quantitative evaluation of peptide bond vibrational intensities in terms of protein secondary structure. Lippert et al.[52] have used Raman data for poly-L-lysine in its α-helical, antiparallel β-pleated sheet, and random coil conformations as standards for frequencies and intensities. Adopting a procedure that is used in studies of protein secondary structure by circular dichroism techniques, they suggest that the Raman spectra of peptide bonds in globular proteins are linear combinations of the intensities due to peptide residues in these three conformational states. A different approach was taken by Lord[6] who postulated a curvilinear relationship between the Amide III frequency and the torsional angle of a given peptide group. A third approach was described by Williams and Dunker[53] who resolved the Amide I band into six components, each representing a different type of secondary structure. Linear combinations of reference spectra were fit to experimental Amide I data to estimate the fractions of residues in each conformation in the unknown sample.

While each of the above schemes has some merit, none has been sufficiently tested to warrant comments on its general validity. Nevertheless, the ability of the technique to give qualitative evaluations of protein secondary structure is clear.

2. Side-Chain Vibrations

In addition to the structural information inherent in the peptide bond vibrations, the modes arising from the amino acid side chains may be useful in evaluation of their conformation and their environment.

Collections of amino spectra are available,[54] but frequency conformational relationships have not, with few exceptions, been well worked out. Lord and co-workers demonstrated the type of empirical approach that seems best to follow. They have compared the Raman spectra of various small globular proteins such as lysozyme and ribonuclease A[54,55] with the spectra of amino acids weighted in the proportions in

LIPID STRUCTURE

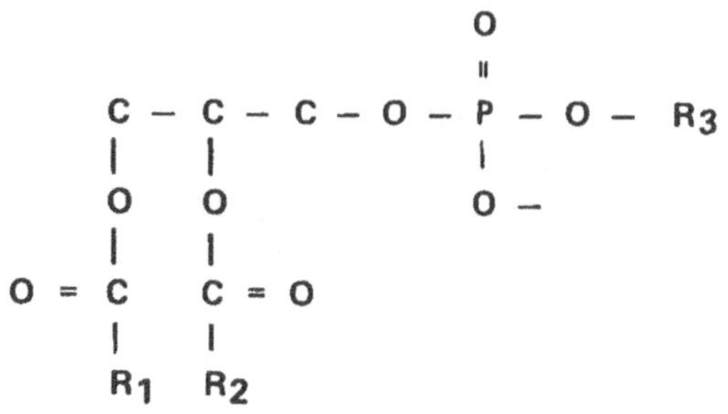

$$O$$
$$\|$$
$$C - C - C - O - P - O - R_3$$

R$_1$, R$_2$ are alkyl chains
C$_{12}$-C$_{24}$, 0-3 C = C
R$_1$ is usually saturated
R$_2$ is unsaturated
R$_3$ is one of several common bases:

If R$_3$ = —C—C—N$^+$(CH$_3$)$_3$, phosphatidylcholine

= —C—C$\diagdown$ $_{COO^-}^{NH_3^+}$, phosphatidylserine

= —C—C—NH$_3^+$, phosphatidylethanolamine

FIGURE 3.   Typical phospholipid molecules found in biological membranes. Note that each molecule has a polar head group and a hydrophobic acyl chain. In aqueous media, two molecules line up so that their hydrocarbon chains are in apposition to each other, thus leading to the spontaneous formation of a phospholipid bilayer.

which they are found in the native enzyme. If the side-chain vibrations in the enzyme compared with the amino acid differ in frequency or intensity, the inference is that the conformation or structure of the residue differs in the two cases. As an example, a doublet in the spectrum of tyrosine at 850 and 830 cm^{-1} has been found to give patterns of relative intensity which are sensitive to the nature of the hydrogen bonding of the tyrosine hydroxyl function or to its state of ionization.[56] In addition, changes in the local geometry around the disulfide C—S—S—C link produce significant changes in the C—S and S—S stretching vibrations in the 500 to 650 cm^{-1} regions of the spectrum. The exact correlation between the spectrum and the structure has been controversial. Sugeta et al.,[57] in a study of dialkyl disulfides, have correlated the S—S frequency with the C—S dihedral angle. This correlation was called into question by Van Wart and Scheraga.[58]

B. Phospholipid Structure

The most common lipid components in biological membranes are phospholipids, and the Raman spectra of biological membranes are dominated by the lipid bands. Typical structures of commonly found phospholipid molecules are shown in Figure 3.

The Raman spectra are dominated by bands arising from the lipid acyl chains, although features arising from the polar head groups are also noted. The large body of literature available on the vibrational spectra of paraffins has facilitated analysis of phospholipid spectra. In an early study Schaufele[59] demonstrated the sensitivity of the

Raman spectrum to changes in the physical state of alkanes having formulas $CH_3(CH_2)_nCH_3$ with n = 13 to 36. In addition, the spectrum of polyethylene, which can be represented as a chain containing an infinite number of methylene groups, has been analyzed in detail,[60,61] so that the origin of the Raman spectrum of the ordered (all-trans) form of that molecule is well understood.

The general methods used in the analysis of polymer spectra have been summarized by Peticolas.[62] Two general types of oscillations are found to exist. Motions which occur perpendicular to the long-chain axis of the polymer are termed transverse phonons, while collective motions that take place along the long axis of the polymer are the longitudinal acoustical vibrations. These motions (termed LAMs) involve accordion-like stretching and compressions of the polymer chains.

In a general type of treatment, a finite polymer that has x atoms per repeating translational unit and y repeating units, there will be 3xy total vibrations which are divided into 3x frequency branches. For a given frequency (for example, the CH_2 symmetric stretching mode), there will be y possible modes of vibration. Each of these will depend on the relative phases of the nuclear displacements in going from cell to cell. Each mode will be characterized by a phase angle θ such that

$$\Theta = k\pi/y \quad \text{with} \quad k = 0,1,2....y - 1$$

The dispersion curves (frequency vs. phase angle plots) for a polyethylene chain are shown in Figure 4.

From theoretical considerations, it is known that for a polymer of infinite length such as polyethylene, only those vibrations with k = 0 (those with identical nuclear displacements in each unit cell) are allowed in the Raman spectrum. For finite chain lengths, the k = 0 selection rule is not rigorously obeyed, and vibrations with other values of k may be observed.

As shown by Snyder[63] and others,[60] the frequency of any vibration with k = 0 is a sensitive function of the all-trans chain length present in the CH_2 oligomer. In addition, those vibrations with sufficient variation of frequency with phase angle (which permit observation of varying k = 0 modes when conformational alterations take place) will be useful as structural probes in membrane systems. This follows as during the gel-liquid phase transition, the average all-trans chain length of a lipid chain is reduced when the lipids melt.

Modes with flat dispersion curves (such as the CH_2 stretching frequencies in Figure 4) show only small changes in frequency when lipids undergo phase transitions. These are, however, easily measurable by FT-IR (see Chapter 3) and reflect very localized variations in the force field around the molecular groupings involved.[64]

Four regions of the Raman spectrum have been used in structural studies of biomembranes. The first is the C—C stretching modes (1000 to 1150 cm^{-1}). The 1000 to 1150 cm^{-1} region contains C—C stretching motions in which alternate carbon atoms in the acyl chains move in opposite directions along the main axis. These skeletal optical modes are seen in the lipid gel phase as strong Raman bands due the k = 0 modes of an all-trans chain at 1060 and 1128 cm^{-1}. Above the lipid melting temperature (T_m), the higher-frequency mode is reduced in intensity and a new feature appears in the 1080 to 1100 region (depending on the chain length), which arises from uncoupled C—C stretching vibrations of chain segments containing gauche rotamers. Lippert and Peticolas[65] made the first observations of the sensitivity to conformational alteration of the 1130/1100 intensity ratio, and later[66] suggested a quantitative model by which this value can be correlated with the number of all-trans bonds in the chain. Several investigators have used this parameter to monitor the effects of external perturbants such as cholesterol on acyl chain conformation.[67-69] The application of the quantitative

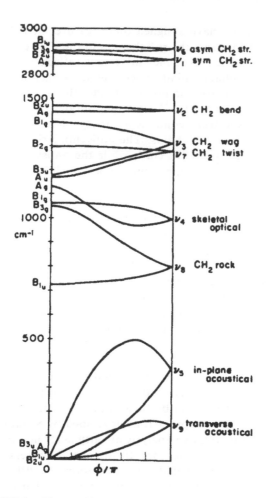

FIGURE 4. Phonon dispersion curve of a polyethylene zigzag (all-trans) chain. The phase angles (abscissa) are given for a $-CH_2-CH$ group. The modes that have been found most useful for Raman spectroscopic studies of model membrane systems are the ν_4 (skeletal optical) and ν_5 (acoustical) vibrations. Other modes with flatter dispersion curves (e.g., the CH_2 stretching modes) have been used in infrared studies of membrane phase alterations. Data are reproduced from Lippert and Peticolas,[65] and had been adapted from the experimental studies of Snyder.[63]

model has been called into question;[70] nevertheless, the qualitative correlation of the intensity parameter with acyl chain order is certain. A more sophisticated interpretation of intensity changes in this spectral region has been given by Pink et al.[71] Initially they calculate, using a statistical mechanical model, the number and types of conformers likely to be present in the hydrocarbon chains. Next, the Raman intensity expected at 1130 cm^{-1} for each state of the chain is computed. The intensity calculations are based upon Theimer's model for polymer Raman intensities[72] coupled with Snyder's normal coordinate calculations.[63] The agreement between theory and experiment is satisfying. However, an obvious drawback of the approach is its inapplicability in situations where the statistical models are inadequate for predictions of the expected conformations (such as in lipid/protein systems).

The second region of the Raman spectrum used in structural studies of biomembranes is the C=O stretching modes (1700 to 1750 cm^{-1}). For most phospholipids, the C=O stretching region contains two spectral features, near 1742 and 1728 cm^{-1}, respectively. Levin and co-workers have assigned these features in PCs to particular acyl chains, the 1742 band arising from the sn1 chain, and the 1728 cm^{-1} band arising from the sn2 chain.[73,74] The observed frequency difference reflects structural inequivalence of the chains, with the sn2 chain initially extending in a direction perpendicular to the sn1, then developing a gauche bend in order to render the two chains parallel.

The third spectral region is the C—H stretching modes (2800 to 3100 cm^{-1}). This region in the vibrational spectrum contains a variety of features arising from C—H stretching modes of the lipid acyl chains. The antisymmetric and symmetric methyl vibrations appear near 2956 and 2872 cm^{-1}, while the antisymmetric and symmetric CH2 stretching bands of the methylene groups appear at 2920 and 2850 cm^{-1} in the IR spectrum. The C—H stretching region of the Raman spectrum contains an additional, very intense feature near 2885 cm^{-1} which depends upon interchain interaction for its sensitivity. The structureless features underlying this band (giving it its intensity) have been analyzed by Snyder et al.,[75] and arise from Fermi resonance between the symmetric CH2 stretch and the appropriate combination of methylene bending modes near 1450 cm^{-1}. The magnitude of this vibrational coupling depends upon interchain interaction, which in turn depends upon the average distance between chains. Therefore, in situations where the acyl chains are not well packed (sonicated vesicles, lipid-protein systems, etc.) even though they may be ordered (much all-trans conformation present), the band at 2880 cm^{-1} will show a marked drop in intensity. This spectral feature has been used to evaluate the effect of sonication and proteins on interchain interaction.[76,77]

The spectral region to be used in studies of biomembranes is the region below 600 cm^{-1}. The spectral region below 600 cm^{-1} contains the LAMs (C—C—C bending modes), the Raman frequencies of which can be fit to a power series in the number of CH2 units in an all-trans chain. This approach, as developed by Schaufele and Shimanouchi[78] has been used[79] for quantitative analysis of fatty acids in a mixture in the C_{12} to C_{24} range. In applications to biomembranes, Brown et al. have reported a Raman study of a dispersion of dipalmitoylphosphatidylethanolamine (DPPE).[80] At temperatures below the gel/liquid-crystal phase transition temperature, a sharp band is noted at 161 cm^{-1}. This is the same position as in solid palmitic acid, and is characteristic of the all-trans chain in DPPE. Above T_m, the mode broadens and shifts to higher frequencies, as expected for the shorter effective all-trans chain length of the many possible gauche structures that can be formed in the liquid-crystalline state. This spectral region is quite difficult to observe, since scattering of the excitation laser light by turbid suspensions is quite intense and the Raman spectrum at small frequency shifts is often obscured. Future studies of this spectral region seem indicated.

C. Reconstituted Lipid-Protein Systems

Native membranes are extremely complex entities. It has been estimated that the erythrocyte membrane contains 150 chemically distinct phospholipid molecules, along with a large number of proteins. Any spectroscopic study of a native system is therefore complicated by the fact that all species contribute to the measured spectral parameter. To simplify the problems of spectral interpretation, many types of biophysical studies have been accomplished with reconstituted systems. Individual integral membrane proteins are isolated, separated from any tightly bound native lipids, and reinserted into a controlled phospholipid environment of the experimenter's choosing. Such preparations have the advantage that the effect of a single type of lipid species on (for example) the function of a membrane-bound enzyme, may be correlated with structural alterations as seen in Raman spectroscopic experiments.

A variety of Raman spectroscopic studies have appeared along these general lines. The interaction of lysozyme (not an integral membrane protein) with liposomes formed from a mixture of the phospholipids 1,2-dipalmitoyl-L-phosphatidic acid/1,2 dimyris-toyl-L-phosphatidylcholine was studied by Lippert et al.[81] At temperatures <27°C, interaction of protein with lipid slightly increased the amount of helical structure in lysozyme, while above 30°C, considerable β-sheet was formed. The onset of β-structure was accompanied by the introduction of disorder into the lipid side chains. The results were interpreted in terms of both polar interactions between the acidic lipid component at temperatures where the lipids are in the gel state and hydrophobic interactions between lipid and protein at higher temperatures. Lis et al.[82] examined the effect of the ionophores alamethicin and valinomycin on phospholipid organization. Both peptides tended to fluidize dimyristoyl lecithin but produced little alteration in the chain conformation of dipalmitoyl lecithin (DPPC). Detailed temperature studies were not reported. In contrast, Verma and Wallach[83] suggested that the peptide mellitin tends to rigidify DPPC multibilayers. In a more thorough investigation of the same system,[84] Levin and co-workers demonstrated two melting events when the peptide was inserted into dimyristoylphosphatidylcholine (DMPC). The primary gel/liquid-crystal transition is depressed when the peptide concentration is increased. The concentration-independent higher-temperature transition is associated with the melting of lipids present at the lipid-polypeptide interface. It was estimated that about six lipid molecules fall into this category.

Along a similar vein, Mendelsohn and co-workers[85,86] studied the phospholipid chain conformation in complexes of the polypeptide hormone glucagon with DMPC and DPPC. The primary effect of protein appeared to be the disordering of the lipid prior to the main chain melt. Above the main transition temperature, little change in order was noted. In addition, the C−H spectral parameters indicated that lateral interactions between lipid molecules were disrupted by protein. Such alterations are indicative of strong hydrophobic interactions between lipid and protein. As a model of electrostatic interactions that occur between peripheral membrane proteins and lipids at the bilayer surface, a study of the DPPC/cardiolipin/insulin complex was also undertaken.[86] The primary interaction in this case was between the insulin (positively charged at the pH used, and the cardiolipin (negatively charged). The electrostatic neutralization was propogated to the acyl chains of the lipids, where an ordering and rigidification was noted. In addition, the ternary complex underwent a broad melting event not seen in the binary lipid mixture.

Lis et al.[87] examined the effect on lipid structure of a variety of intrinsic and extrinsic membrane proteins. Human fibrinogen and bovine serum albumin induced intensity changes in the C−H stretching region of DMPC, but no detailed structural analysis was undertaken.

In a study of a native membrane protein, Curatolo et al. obtained Raman spectra of myelin apoprotein proteolipid reconstituted with both DMPC and egg lecithin.[88] Spectra of the C−H stretching region showed that the DMPC acyl chains possessed some residual "gel" character at temperatures as high as 18°C, above T_m for this phospholipid. When the apoprotein was combined with egg lecithin, the main chain melt still occurred below 0°C, but a new structural transition was observed at 12°C. The Raman and calorimetric results presented suggested that some of the phospholipid was sequestered by the protein into regions that can undergo a cooperative thermal transition higher than normal, and presumably consists of saturated chains.

As a final demonstration of the use of Raman spectroscopy in reconstitution experiments, Taraschi and Mendelsohn[77] worked with an integral membrane protein (glycophorin) isolated from human erythrocye membranes. The molecule was inserted into liposomes consisting of DPPC or its acyl chain perdeuterated analog (DPPC-d_{62}). At lipid/protein mole ratios of 125:1, a broad melting event much reduced in temperature from the pure lipid was noted. It was apparent that this number of lipid molecules had their phase transition drastically altered by protein.

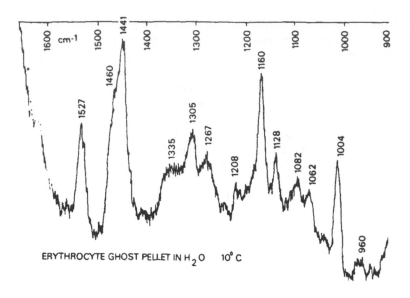

FIGURE 5.   Raman spectrum of red blood cell membranes (actually, the hemo-globin has been removed so that these fragments are termed erythrocyte ghosts) at 10°C. A strong carotenoid feature is noted near 1527 cm⁻¹. (From Lippert, J. L. et al., *Biochim. Biophys. Acta,* 382, 51, 1975. With permission.)

D. Applications to Native Membranes

As indicated in the preceding text, native membranes present several difficulties for Raman spectroscopic studies. The occurrence of fluorescent constituents is wide-spread. Signals from all species in the membrane are obtained so that the contribution of a particular component to the spectrum is uncertain. Finally, as pointed out by Forrest,[89] hysteresis in temperature studies may be common, so that the thermal history of the sample must be considered as an experimental variable.

To date few studies have appeared on native membranes. It seems appropriate to suggest that the conclusions drawn from this body of work must be considered tentative until a greater body of experimental evidence is amassed.

The first report of Raman scattering from a native membrane appears to have been due to Bulkin,[90] who observed one or two Raman lines from a suspension of hemoglobin-depleted erythrocyte membranes (erythrocyte ghosts). No useful structural information was gained. In a later, more detailed study of the same system, Lippert et al.[91] obtained beautiful spectra (see Figure 5) of the same membrane (heavily washed, spectrin depleted) in aqueous suspension. Amide I, Amide I', and the Amide III vibrational patterns were used to assess the protein conformation at 40 to 55% α-helix with little or no pleated sheet (these numbers are averages over all the proteins present). Strong features at 1527 cm⁻¹ and 1160 cm⁻¹ in Figure 5 have now been identified as resonance-enhanced vibrations of β-carotene. The presence of this molecule was confirmed subsequently by Verma and Wallach[92] who further attempted to use the C−C and C═C stretching bands of the carotenoid as a structural probe.

In addition to information about the protein component, Lippert et al.[91] concluded that the phospholipids had about 60% of their hydrocarbon chains in the all-trans conformation. In view of more recent developments concerning spectra structure correlations in the lipid C−C bands used for the analysis, the number must be regarded with caution. Verma and Wallach further studied the temperature dependence of the C−H stretching region and observed the main structural transition at about −8°C.[93] An irreversible conformational change at 40°C was attributed to alterations at sites of lipid-protein interaction.

A different system, in this instance containing a single dominant species of membrane protein, was investigated by Rothschild et al.[94] in their studies of the visual protein opsin. Membranes from calf were isolated and their Raman spectra acquired with 5145 Å excitation, after bleaching of the chromophore (retinylidenelysine) to eliminate the resonance Raman effect. The authors concluded that opsin has predominantly a helical structure with little if any pleated sheet present. A substantial fraction of the tyrosines in the opsin appears to be strongly hydrogen bonded (the intensity ratio 850/830 <1), while the phospholipid hydrocarbon chains were in a fluid state.

An evolving element of the research tactics in this area appears to be the use of native membranes which contain a dominant protein species. An example of a membrane-bound enzyme widely used for biophysical studies is the Ca-Mg-dependent ATPase which comprises 80% of the total protein in rabbit sarcoplasmic reticulum. Preliminary studies of SR membranes were undertaken by Milanovich et al.[95] The phospholipids were found to be in a more fluid state than in erythrocyte membranes. A more detailed study was undertaken by Lippert and co-workers.[96] In their report of the temperature dependence of the Raman spectrum of the SR, no discontinuities were noted in the 10 to 37°C range. The protein conformation was predominantly α-helix. The protein itself appeared to undergo a conformational change at 15 to 18°C which involved (1) increased α-helical content and (2) removal of a tryptophan residue from an aqueous environment. This change coincided with a change in the slope of the Arrhenius plots of ATP hydrolysis activity.

E. Resonance Raman Studies of Membrane Structure

The bulk of the resonance Raman investigations in the membrane area have dealt with structure-function relationships of the visual pigment rhodopsin. This important topic has been reviewed several times recently, and the interested reader is referred to the literature citations.[10]

Additional possibilities exist for the study of the properties and functions of natural membranes using resonance Raman probes, either of an intrinsic or extrinsic nature. The former was demonstrated feasible by Koyama et al.,[97] who used as a probe molecule, the carotenoids which are a natural constituent in the membranes of the photosynthetic bacteria. Changes in the resonance Raman spectra of the carotenoids in these membranes are useful probes of membrane potential. Alteration in electrial potential induces small changes in the carotenoid visible absorption spectra; these may be effectively amplified by suitable choice of excitation frequency in resonance Raman studies. The intensity ratio of the C=C to the C−C stretching modes is the parameter of interest. Another example of the use of a polyene resonance Raman spectrum involves the extrinsic probe amphotericin B, a polyene antibiotic containing seven conjugated double bonds. The antibiotic was inserted into a model membrane system. It was possible[98] to monitor changes both in the resonance Raman spectrum of the antibiotic and the normal Raman spectrum of the lipid as a function of temperature.

In a final example,[99] the dye quinaldine red has been added as an extrinsic probe to the cells of the bacterium *Streptococcus faecalis*. The spectrum of the free dye is altered upon binding to the membrane of resting cells and is further modified when the membranes are energized.

REFERENCES

1. Durig, J. R. and Harris, W. C., Raman spectroscopy, in, *Physical Methods of Chemistry*, Part III B. Weissberger, A. and Rossiter, B. W., Eds., Interscience, New York, 1972, Chapt. 2.
2. Fawcett, V. and Long, D. A., Vibrational spectroscopy of macromolecules, *Mol. Spectrosc.*, 1, 1973.
3. Szymanski, H. A., Ed., *Raman Spectroscopy*, Vols. 1 and 2, Plenum Press, New York, 1967, 1970.
4. Yu, N.-T., Raman spectroscopy: a conformational probe in biochemistry, *CRC Crit. Rev. Biochem.*, 4, 229, 1977.
5. Spiro, T. G. and Stein, P., Resonance effects in vibrational scattering from complex molecules, *Annu. Rev. Phys. Chem.*, 28, 501, 1977.
6. Lord, R. C., Strategy and tactics in the Raman spectroscopy of biomolecules, *Appl. Spectrosc.*, 31, 187, 1977.
7. Herzberg, G., *Infrared and Raman Spectra*, Van Norstrand, New York, 1945.
8. Placzek, G., Rayleigh-Streuung und Raman-Effekt, in, *Handbuch der Radiologie*, Vol. 6, Marx, E., Ed., Akademische Verlag, Leipzig, 1934, 205. (English translation (UCRL-Trans-526 [L]) available from Clearing House for Federal Scientific and Technical Information, Springfield, Va., 22151.)
9. Spiro, T. G., Biological applications of resonance Raman spectroscopy: haem proteins, *Proc. R. Soc. London Ser. A*, 845, 89, 1975.
10. Carey, P. R., *Biochemical Applications of Raman and Resonance Raman Spectroscopies*, Academic Press, New York, 1982.
11. Peticolas, W. L., Applications of Raman spectroscopy to biological macromolecules, *Biochimie*, 57, 417, 1975.
12. Koningstein, J. A., *Introduction to the Theory of the Raman Effect*, Reidel, Dortrecht, 1972.
13. Long, D. A., *Raman Spectroscopy*, McGraw-Hill, New York, 1977.
14. Hartmann, K. A., Lord, R. C., and Thomas, G. J., Jr., Structural studies on nucleic acids and polynucleotides by infrared and Raman spectroscopy, in *Physical Chemistry of Nucleic Acids*, Vol. 2, Duchesne, J., Ed., Academic Press, New York, 1973, chap. 10.
15. Carey, P. R., Resonance Raman spectroscopy in biochemistry and biology, *Q. Rev. Biophys.*, 11, 309, 1978.
16. Spiro, T. G. and Strekas, T. C., Resonance Raman spectra of hemoglobin and cytochrome c: inverse polarization and vibronic scattering, *Proc. Natl. Acad. Sci. U.S.A.*, 69, 2622, 1972.
17. Storer, A. C., Murphy, W. F., and Carey, P. R., The use of resonance Raman spectroscopy to monitor catalytically important bonds during enzymic catalysis, *J. Biol. Chem.*, 254, 3163, 1979.
18. Spiro, T. G., Resonance Raman spectroscopy: a new structure probe for biological chemophores, *Acc. Chem. Res.*, 7, 339, 1974.
19. Bellamy, L. J., *Infrared Spectra of Complex Molecules*, 2nd ed., Methuen, London, 1958.
20. Bellamy, L. J., *Advances in Infrared Group Frequencies*, Methuen, London, 1968.
21. Stein, P., Burke, J. M., and Spiro, T. G., Structural interpretation of heme protein resonance Raman frequencies. Preliminary normal coordinate analysis results, *J. Am. Chem. Soc.*, 97, 2304, 1975.
22. Ziegler, L. D., Strommen, D. P., Hudson, B. S., and Peticolas, W. L., Ultraviolet resonance Raman studies of mononucleotides using 213 nm and 266 nm excitation, in, *Raman Spectroscopy, Linear and Nonlinear*, Lascombe, J. and Huong, P. V., Eds., John Wiley & Sons, New York, 1982.
23. Nishimura, Y., Hirakawa, A. Y., Tsuboi, M., and Nishimura, S., Raman spectra of transfer RNAs with ultraviolet lasers, *Nature (London)*, 260, 173, 1976.
24. Sugawara, Y., Harada, I., Matsuura, H., and Shimanouchi, T., Preresonance Raman studies of poly(L-lysine), poly(L-glutamic acid), and deuterated N-methylacetamides, *Biopolymers*, 17, 1405, 1978.
25. Maker, P. D. and Terhune, R. W., Study of optical effects due to an induced polarization of third order in the electric field strength, *Phys. Rev.*, 137A, 801, 1965.
26. Hudson, B. S., New laser techniques for biophysical studies, *Annu. Rev. Biophys. Bioeng.*, 6, 135, 1977.
27. Nestor, J., Spiro, T. G., and Klauminzer, G., Coherent Anti-Stokes Raman Scattering (CARS) spectra with resonance enhancement of cytochrome C. and vitamin B_{12} in dilute aqueous solution, *Proc. Natl. Acad. Sci. U.S.A.*, 73, 3329, 1976.
28. Irwin, R. M., Visser, A. J., Lee, J., Carreira, L. A., Protein-ligand interactions in lumazine protein and in desulfovibrio flavodoxins from resonance coherent anti-Stokes Raman spectra, *Biochemistry*, 19, 4639, 1980.
29. Kataoka, H., Kamisuki, T., Adachi, Y., and Maeda, S., Resonance CARS and CSRS of excited pyrene, in *Raman Spectroscopy, Linear and Nonlinear*, Lascome, J. and Huong, P. V., Eds., John Wiley & Sons, New York, 1982.
30. Bridoux, M., Recent advances in pulsed Raman spectroscopy, *Proc. 6th Int. Conf. Raman Spectroscopy*, Vol. 1, Schmid, E. D., Krishnan, R. S., Kiefer, W., and Schrotter, H. W., Eds., Heyden and Son, Ltd., London, 1978.

31. Delhaye, M., Defontaine, A., Bridoux, M., Hug, W., and Da Silva, E., A new multi-channel Raman spectrometer, in *Raman Spectroscopy, Linear and Nonlinear,* Lascombe, J. and Huong, P. V., Eds., John Wiley & Sons, New York, 1982.

32. Buiteveld, H., De Mul, F. F. M., Greve, J., and Mud, J., Raman spectrometry of microscopical particles in human lung tissue, in *Raman Spectroscopy, Linear and Nonlinear,* Lascombe, J. and Huong, P. V., Eds., John Wiley & Sons, New York, 1982.

33. Burstein, E. and Chen, C. Y., Raman scattering by molecules adsorbed at metal surfaces. The role of surface roughness, in *Proc. 7th Int. Conf. Raman Spectroscopy,* Murphy, W. F., Ed., North Holland, Amsterdam, 1980.

34. Wood, T. H. and Klein, M. L., Studies of the mechanism of enhanced Raman scattering in ultrahigh vacuum, *Solid State Commun.,* 35, 263, 1980.

35. Creighton, J. A., Blatchford, C. G., and Albrecht, M. G., Plasma resonance enhancement of Raman scattering by pyridine adsorbed on silver or gold sol particles of size comparable to the excitation wavelength, *J. Chem. Soc. Faraday Trans. 2,* 75, 790, 1979.

36. Moskovitz, M., Enhanced Raman scattering by molecules adsorbed on electrodes — a theoretical model, *Solid State Commun.,* 32, 59, 1979.

37. Van-Duyne, R. P., in *Chemical and Biological Applications of Lasers,* Vol. 5, Moore, C. B., Ed., Academic Press, New York, 1979, chap. 4.

38. Sequaris, J.-M., Koglin, E., Valenta, P., and Nurnberg, H. W., Studies on the interfacial behaviour of nucleic acids by SERS, in, *Raman Spectroscopy, Linear and Nonlinear,* Lascombe, J. and Houng, P. V., Eds., John Wiley & Sons, New York, 1982.

39. Zubay, G., *Biochemistry,* Addison-Wesley, Reading, Mass., 1983.

40. Miyazawa, T., Shimanouchi, T., and Mizushima, S., Normal vibrations of N-methyl acetamide, *J. Chem. Phys.,* 29, 611, 1958.

41. Miyazawa, T. and Blout, E. R., The infrared spectra of polypeptides in various conformations. Amide I and II bands, *J. Am. Chem. Soc.,* 83, 712, 1961.

42. Koenig, J. L., Raman spectroscopy of biological molecules. A review, *J. Polymer Sci., Part D,* 6, 59, 1972.

43. Frushour, B. G., and Koenig, J. L., Raman spectroscopic study of mechanically deformed poly-L-alanine, *Biopolymers,* 13, 455, 1974.

44. Yu, T.-J., Lippert, J. L., and Peticolas, W. L., Laser Raman studies of conformational variations of poly-L-lysine, *Biopolymers,* 12, 2161, 1973.

45. Bandekar, J., Evans, D. J., Krimm, S., Leach, S. J., Lee, S., McQuie, J. R., Minasian, E., Nemethy, G., Pottle, M. S., and Scheraga, H. A., Conformations of cyclo-L-alanyl-aminocaproyl and of cyclo-L-analyl-D-analyl-aminocaproyl. Cyclized dipeptide models for specific types of beta bends, *Int. J. Pept. Protein Res.,* 19, 187, 1982.

46. Yu, N.-T. and Liu, C. S., Laser Raman spectra of cyrstalline and aqueous glucagon, *J. Am. Chem. Soc.,* 94, 5127, 1972.

47. Chen, M. C., Lord, R. C., and Mendelsohn, R., Laser-excited Raman spectroscopy of biomolecules. V. Conformational changes associated with the chemical denaturation of lysozyme, *J. Am. Chem. Soc.,* 96, 3038, 1974.

48. Chen, M. C. and Lord, R. C., Laser-Raman spectroscopic studies of the thermal unfolding of ribonuclease A, *Biochemistry,* 15, 1889, 1976.

49. Dunker, A. K., Williams, R. W., Gaber, B. P., and Peticolas, W. L., Laser Raman studies of lipid disordering by the β-proteins of fd phage, *Biochim. Biophys. Acta,* 553, 351, 1979.

50. Yu, N.-T., Kuck, J. F. R., Jr., and Askren, C. C., Laser Raman spectroscopy of the lens, in situ measured in an anesthetized rabbit, *Curr. Eye Res.,* 1(10), 615, 1981—82.

51. Mizuno, A., Ozaki, Y., Kamada, Y., Miyazaki, H., Itoh, K., and Iriyamo, K., Direct measurement of Raman spectra of intact lens in a whole eyeball, *Curr. Eye Res.,* 1(10), 609, 1981—82.

52. Lippert, J. L., Tyminski, D., and Desmeules, P. J., Determination of the secondary structure of proteins by laser Raman spectroscopy, *J. Am. Chem Soc.,* 98, 7075, 1976.

53. Williams, R. W. and Dunker, A. K., Determination of the secondary structure of proteins from the Amide I band of the laser Raman spectrum, *J. Mol. Biol.,* 152(4), 783, 1981.

54. Lord, R. C. and Yu, N.-T., Laser-Raman spectroscopy of biomolecules. I. Native lysozyme and its constituent amino acids, *J. Mol. Biol.,* 51, 509, 1970.

55. Lord, R. C. and Yu, N.-T., Laser-Raman spectroscopy of biomolecules. II. Native ribonuclease and alpha-chymotrypsin, *J. Mol. Biol.,* 52, 203, 1970.

56. Siamwiza, M. N., Lord, R. C., Chen, M. C., Tokamatsu, T., Harada, I., Matsuura, H., and Shimanouchi, T., Interpretation of the doublet at 850 and 830 cm^{-1} in the Raman spectra of tyrosyl residues in proteins and certain model compounds, *Biochemistry,* 14, 4870, 1975.

57. Sugeta, H., Go, A., and Miyazawa, T., Vibrational spectra and molecular conformations of dialkyl disulphides, *Bull. Chem. Soc. Jpn.,* 46, 3407, 1973.

58. Van Wart, H. E. and Scheraga, H. A., Raman spectra of strained disulphides: effect of rotation about S–S bonds on S–S stretching frequencies, *J. Phys. Chem.*, 80, 1823, 1976.
59. Schaufele, R. F., Chain shortening in polymethylene liquids, *J. Chem. Phys.*, 49, 4168, 1968.
60. Tasumi, M. and Krimm, S., Crystal vibrations of polyethylene, *J. Chem. Phys.*, 46, 755, 1967.
61. Abbate, S., Zerbi, G., and Wunder, S. L., Fermi resonances and vibrational spectra of crystalline and amorphous polymethylene chains, *J. Phys. Chem.*, 86, 3140, 1982.
62. Peticolas, W. L., Mean-square amplitudes of the longitudinal vibrations of helical polymers, *Biopolymers*, 18(4), 747, 1979.
63. Snyder, R. G., Vibrational study of the chain conformation of the liquid n-paraffins and molten polyethylene, *J. Chem. Phys.*, 47, 1316, 1967.
64. Snyder, R. G., Strauss, H. L., and Ellinger, C. A., C–H stretching modes and the structure of n-alkyl chains. L. Long disordered chains, *J. Phys. Chem.* 86, 5145, 1982.
65. Lippert, J. L. and Peticolas, W. L., Laser-Raman investigation of the effect of cholesterol on conformational changes in dipalmitoyl lecithin multilayers, *Proc. Natl. Acad. Sci. U.S.A.*, 68, 1572, 1972.
66. Lippert, J. L. and Peticolas, W. L., Raman-active vibrations in long chain fatty acids and phospholipid sonicates, *Biochim. Biophys. Acta*, 282, 8, 1972.
67. Mendelsohn, R., Laser-Raman spectroscopic study of egg lecithin and egg lecithin-cholesterol mixtures, *Biochim. Biophys. Acta*, 290, 15, 1972.
68. Lieb, W. L. and Mendelsohn, R., Do clinical levels of general anesthetics affect lipid bilayers? Evidence from Raman scattering, *Biochim. Biophys. Acta*, 688, 388, 1982.
69. Lis, L. -J., Kauffman, J. W., and Shriver, D. F., Effect of ions on phospholipid layer structures as indicated by Raman spectroscopy, *Biochim. Biophys. Acta*, 406, 453, 1975.
70. Karvaly, B. and Loschilova, E., Comments on the quantitative interpretation of biomembrane structure by Raman spectroscopy, *Biochim. Biophys. Acta*, 470, 492, 1977.
71. Pink, D. A., Green, T. J., and Chapman, D., Raman scattering in bilayers of saturated phosphatidylcholines. Experiment and theory, *Biochemistry*, 19, 349, 1980.
72. Theimer, O. H., Raman and infrared intensities in the vibrational spectra of hydrocarbons. I. Skeletal vibrations of straight zigzag chains, *J. Chem. Phys.*, 27, 408, 1957.
73. Mushayakarara, E., Albon, N., and Levin, I., Effect of water on the molecular structure of a phosphatidylcholine hydrate. Raman spectroscopic analysis of the phosphate carbonyl and carbon-hydrogen stretching mode regions of 1,2 dipalmitoylphosphatidylcholine dihydrate, *Biochim. Biophys. Acta*, 686, 153, 1982.
74. Levin, I. W., Mushayakarara, E., and Bittman, R., Vibrational assignment of the SN-1 and SN-2 chain carbonyl stretching modes of membrane phospholipids, *J. Raman Spectrosc.*, 13, 231, 1982.
75. Snyder, R. G., Hus, S. L., and Krimm, S., Vibrational spectra in the C–H stretching region and the structure of the polymethylene chain, *Spectrochim. Acta*, 34A, 395, 1978.
76. Spiker, R. C. and Levin, I. W., Effect of bilayer curvature on vibrational Raman spectroscopic behavior of phospholipid-water assemblies, *Biochim. Biophys. Acta*, 455, 560, 1976.
77. Taraschi, T. -F. and Mendelsohn, R., Lipid-protein interaction in the glycophorin-dipalmitoylphosphatidyl choline system. Raman spectroscopic investigation, *Proc. Natl. Acad. Sci. U.S.A.*, 77, 2362, 1980.
78. Schaufele, R. F. and Shimanouchi, T., Longitudinal acoustical vibrations of finite polymethylene chains, *J. Chem. Phys.*, 47, 3605, 1967.
79. Warren, C. H. and Hooper, D. L., Chain length determination of fatty acids by Raman spectroscopy, *Can. J. Chem.*, 51, 3901, 1973.
80. Brown, K. G., Peticolas, W. L., and Brown, E., Raman studies of conformational changes in model membrane systems, *Biochem. Biophys. Res. Commun.*, 54, 358, 1973.
81. Lippert, J. L., Lindsay, R. M., and Schultz, R., Laser-Raman investigation of lysozyme-phospholipid interactions, *Biochim. Biophys. Acta*, 599, 32, 1980.
82. Lis, L. J., Kauffman, J. W., and Shriver, D. F., Raman spectroscopic detection and examination of the interaction of amino acids, polypeptides and proteins with the phosphatidylcholine lamellar structure, *Biochim. Biophys. Acta*, 443, 331, 1976.
83. Verma, S. P. and Wallach, D. F. H., Effect of melittin on thermotropic lipid state transitions in phosphatidylcholine liposomes, *Biochim. Biophys. Acta*, 426, 616, 1976.
84. Levin, I. W., Lavialle, F., and Mollay, C., Comparative effects of melittin and its hydrophobic and hydrophilic fragments on bilayer organization by Raman spectroscopy, *Biophys. J.*, 37, 339, 1982.
85. Taraschi, T. and Mendelsohn, R., Raman scattering from glucagon-dimyristoyl lecithin complexes. A model system for serum lipoproteins, *J. Am. Chem. Soc.*, 101, 1050, 1979.
86. Mendelsohn, R., Taraschi, T., and Von Holten, R. W., Raman spectroscopic studies of the dipalmitoylphosphatidylcholine/glucagon and the dipalmitoylphosphatidycholine/cardiolipin/insulin systems, *J. Raman Spectrosc.*, 8, 279, 1979.

87. Lis, L. J., Goheen, S. C., Kauffman, J. W., and Shriver, D. F., Laser-Raman spectroscopy of lipid-protein systems. Differences in the effect of intrinsic and extrinsic proteins on the phosphatidylcholine Raman spectrum, *Biochim. Biophys. Acta*, 443, 331, 1976.

88. Curatolo, W., Verma, S. P., Sakura, J. D., Small, D. M., Shipley, G. G., and Wallach, D. F. H., Structural effects of myelin proteolipid apoprotein on phospholipids. A Raman spectroscopic study, *Biochemistry*, 17, 1802, 1978.

89. Forrest, G., Raman spectroscopy of the milk globule membrane and triglycerides, *Chem. Phys. Lipids*, 21, 237, 1978.

90. Bulkin, B., Raman spectroscopic study of human erythrocyte membranes, *Biochim. Biophys. Acta*, 274, 649, 1972.

91. Lippert, J. L., Gorczyca, L. F., and Meiklejohn, G., A laser-Raman investigation of phospholipid and protein configurations in hemoglobin-free erythrocyte ghosts, *Biochim. Biophys. Acta*, 382, 51, 1975.

92. Verma, S. P. and Wallach, D. F. H., Carotenoids as Raman active probes of erythrocyte membrane structure, *Biochim. Biophys. Acta*, 426, 616, 1976.

93. Verma, S. P. and Wallach, D. F. H., Multiple thermotropic state transitions in erythrocyte membranes. A laser Raman study of the C−H stretching and acoustical regions, *Biochim. Biophys. Acta*, 436, 307, 1976.

94. Rothschild, K. H., Andrew, J. R., DeGrip, W. J., and Stanley, H. E., Opsin structure probed by Raman spectroscopy of photo-receptor membranes, *Science*, 191, 1176, 1976.

95. Milanovich, F. P., Yeh, Y., Baskin, R. J., and Harney, R. C., Raman spectroscpic investigations of sarcoplasmic reticulum membranes, *Biochim. Biophys. Acta*, 419, 243, 1976.

96. Lippert, J. R., Lindsay, R. M., and Schultz, R., Laser Raman characterizations of conformational changes in sarcoplasmic reticulum induced by temperature, Ca²⁺, and Mg²⁺, *J. Biol. Chem.*, 256, 12411, 1981.

97. Koyama, Y., Long, R. A., Martin, W. G., and Carey, P. R., The resonance Raman spectrum of carotenoids as an intrinsic probe for membrane potential. Oscillatory changes in the spectrum of neurosporene in the chromatophores of *Rhodopseudomonas spaeroides*, *Biochim. Biophys. Acta*, 548, 153, 1979.

98. Bunow, M. R. and Levin, I. W., Vibrational Raman spectra of lipid systems containing amphotericin B, *Biochim. Biophys. Acta*, 464, 202, 1977.

99. Koyama, Y., Carey, P. R., Long, R. A., Martin, W. G., and Schneider, H., A resonance Raman and electronic absorption probe of membrane energization: quinaldine red in cells of *Streptococcus faecalis*, *J. Biol. Chem.*, 254, 10276, 1979.

Chapter 8

BIOLOGICAL APPLICATIONS OF TOTAL INTERNAL REFLECTION FLUORESCENCE

Seth A. Darst and Channing R. Robertson

TABLE OF CONTENTS

I. INTRODUCTION

The interactions of macromolecules with solid surfaces are complex and not well characterized but are of considerable importance in a number of biological applications. For example, the first step in the complicated process of thrombogenesis, either on natural or foreign surfaces in contact with blood, is recognized to be the adsorption of plasma proteins.[1-3] In addition, many laboratory analytical assays and separation techniques rely on the adsorption of proteins to solid surfaces. These include adsorption chromatography for the separation of proteins[4] and immunoadsorbent assays.[5] Furthermore, the surface diffusion of nonspecifically adsorbed macromolecules is thought to enhance binding with specific cell-surface receptors.[6] Also, the interactions between macromolecules associated with cell membranes and solid surfaces or with other membranes are of obvious importance throughout the field of cellular biology. These are just a few selected instances where macromolecule/surface interactions play a predominant role.

Total internal reflection fluorescence (TIRF)[7-9] is an optical technique especially well suited to the study of macromolecule/surface interactions. In particular, TIRF can be used to continuously monitor fluorescent macromolecules either close to (within about 2000 Å or less) or directly adsorbed to a solid surface. This chapter is a brief overview of some selected examples of recent applications of TIRF to the study of protein adsorption to solid surfaces. The chapter begins with a brief overview of theoretical aspects related to TIRF.

II. TIRF THEORY

TIRF theory is discussed in considerable detail elsewhere.[7] Only the relevant theoretical background will be given here. Consider two transparent media having indices of refraction, n_1 and n_2, with a light beam directed at the interface separating them (Figure 1). If $n_2 < n_1$, and the incident angle $\alpha > \sin^{-1}(n_2/n_1)$, the beam will totally internally reflect at the interface in accordance with Snell's law. An evanescent wave is generated at the point of reflection. The evanescent wave is a standing wave with an amplitude that exponentially decays as it penetrates into medium 2. The depth of penetration, d_p, the distance at which the intensity of the evanescent field decays to e^{-1} of its interfacial value, depends on the incident wavelength, λ, the angle of incidence, α, and the refractive index ratio, and is given by

$$d_p = \frac{\lambda}{2\pi[\sin^2\alpha - (n_2/n_1)^2]^{1/2}}$$

An experimental apparatus can be designed such that the penetration depth of the evanescent wave is 2000 Å or less. If the interface between media 1 and 2 is the surface of interest, and medium 2 is a solution containing fluorescent macromolecules, the interactions of the macromolecules with the surface of medium 1 can be monitored by measuring the fluorescence emitted from molecules excited in the evanescent field.

III. APPLICATIONS

As mentioned in the introduction, an understanding of the mechanisms associated with the adsorption of proteins on a foreign surface is of importance in a number of biological applications. Factors recognized to be fundamentally relevant to protein/surface interactions include adsorption, desorption, and diffusion kinetics, chemical equilibrium between surface-adsorbed proteins and free solution proteins, effects of

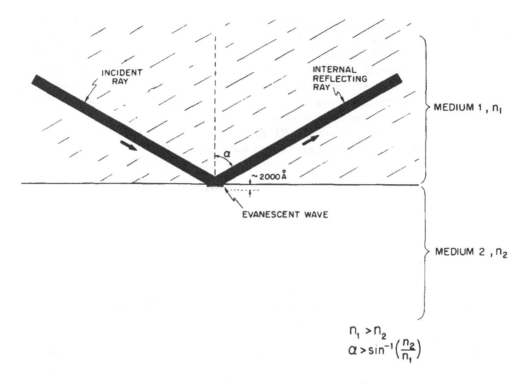

FIGURE 1. Geometry of total internal reflection.[21]

the hydrodynamic environment on the adsorption process, and the conformation of the adsorbed protein layer. An additional factor is the dynamic characteristics of the adsorbed layer, as conformational changes may continue after adsorption occurs.

Techniques that have been used to investigate one or more of these factors include multiple attenuated internal reflection (MAIR) infrared spectroscopy,[1,10] ellipsometry,[1,11] solution depletion,[12] wet chemical assay techniques,[13] radiolabeling,[14-18] circular dichroism,[19] and Fourier transform infrared spectroscopy-attenuated total reflection (FTIR-ATR).[20] (Also refer to Chapters 2 and 3 of this volume.) The following characteristics render TIRF unique from these other methods:

1.   TIRF can be used to continuously and noninvasively monitor the adsorption of fluorescent macromolecules from solutions that are stagnant or flowing.[21]
2.   Since proteins can be specifically labeled, TIRF can be used to study competitive adsorption from multiple protein solutions.[22-24]
3.   TIRF can provide the high sensitivity characteristic of fluorescence techniques.[9]
4.   It is possible to link TIRF with other fluorescence techniques such as fluorescence depolarization,[25] fluorescence energy transfer,[26] and fluorescence photobleaching recovery and correlation spectroscopy.[27-29]

Protein adsorption can be investigated with TIRF using either the intrinsic fluorescence of tyrosine and tryptophan residues contained within proteins[30-32] or extrinsic chromophores conjugated to proteins.[21-29,33,34] Intrinsic chromophores have the obvious advantage that fluorescent labeling is not required. Extrinsic chromophores afford higher sensitivity. Extrinsic chromophores also allow multiple protein studies to be performed since only the adsorption of a labeled species can be distinguished when labeled and unlabeled species are mixed. Extrinsic chromophores should be used at low labeling densities since the labels could conceivably perturb the adsorption behavior of proteins. Adsorption experiments conducted with fluorescein isothiocyanate labeled bovine serum albumin at various labeling densities confirmed that fluorescence intensities vary linearly with labeling density.[21] Thus, with this protein/label pair in the labeling density range observed, the labeling procedure does not alter the amount of protein adsorbed. Dynamic light-scattering measurements before and after labeling also demonstrated that, with this protein-label pair, no detectable changes in the protein diffusion coefficient had occurred upon labeling.[21]

Any surface can be studied using TIRF as long as the surface reflects the incident light beam and allows the resultant evanescent wave to penetrate. Surfaces are typically prepared by coating one side of a transparent slide (glass or quartz) with a thin, transparent film of the material of interest. Considerable care must be taken in the preparation of surfaces to minimize the scattering of light by the surface and the subsequent excitation of bulk proteins.

TIRF studies of protein adsorption can be divided into two categories: (1) the quantitative investigation of kinetic, diffusion, and equilibrium characteristics; and (2) the investigation of adsorbed protein conformation.

A. Protein Adsorption Equilibria, Kinetics, and Surface Diffusion

Following introduction of a protein solution into medium 2 (Figure 1), protein will diffuse to and then adsorb on the interface between media 1 and 2. If these proteins contain fluorescent moieties, either intrinsic or extrinsic, that adsorb energy at the wavelength of the incident light, then the adsorbed protein as well as any protein in solution within the evanescent field will fluoresce. Thus, adsorption and desorption rates, and equilibrium surface concentrations can be determined using appropriate calibration techniques. Calibration considerations have been discussed in detail elsewhere[35] and will not be considered here.

The basic optical requirements for such measurements are

1. A light source
2. A means of aligning the incident beam to achieve total internal reflection at the surface/protein solution interface
3. A detector to measure the intensity of fluorescence emitted by adsorbed proteins

Figure 2 shows a typical TIRF optical design used for protein adsorption experiments.

1. Protein Adsorption Equilibria

The TIRF technique is valuable for its ability to monitor both reversibly and irreversibly adsorbed proteins. No other technique has been as useful in this respect. Some of the protein/surface pairs reported in TIRF studies to exhibit both reversibly and irreversibly adsorbed proteins are the following: bovine serum albumin (BSA) and fibrinogen on cross-linked polydimethylsiloxane (PDMS),[22] BSA on quartz,[28] and immunoglobulin and insulin on BSA-coated quartz.[29] Care must be taken to ensure that decreases in fluorescence intensity due to quantum yield changes are not interpreted as protein desorption.[34]

As an example of a system exhibiting multiply sorbed states, an isotherm for BSA adsorption on cross-linked PDMS obtained with TIRF by Lok et al.[22] is shown in Figure 3. This isotherm was best represented by a mathematical model incorporating a Langmuir reversible component and an irreversible component with 26% of the saturated layer irreversibly adsorbed. Exchange experiments were conducted for long periods of time to determine the exchangeable fraction of adsorbed protein. After 100 hr of exchange, it was found that a bulk concentration of 10 mg%, a wall shear rate of 58 sec^{-1}, and a temperature of 37°C, approximately 40% of the adsorbed layer was replaced.

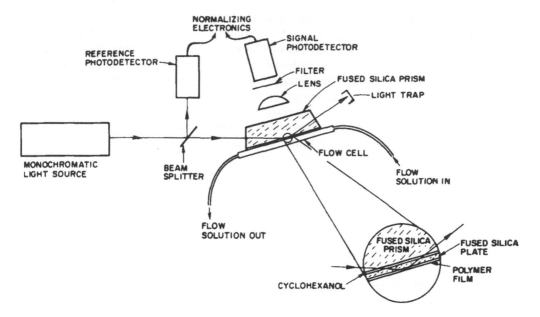

FIGURE 2. Diagram of TIRF apparatus used by Lok et al.[21]

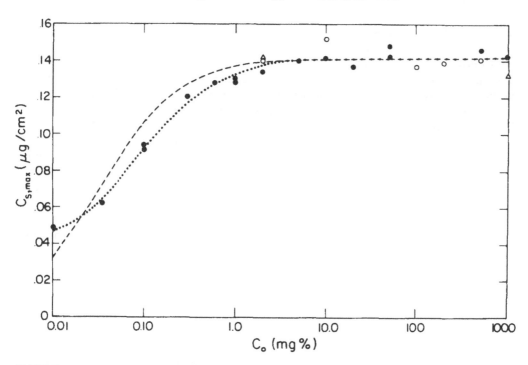

FIGURE 3. BSA/PDMS isotherm. Equilibrium surface concentration, C_s, vs. solution concentration, C_0; closed circles, 100% labeled protein; open circles, 50% labeled protein; triangles, 25% labeled protein; dotted line given by:

$$C_{s,max} = \rho_i\, C_{s,max}^* + \frac{(1 - \rho_i)C_{s,max}^*\, KC_0}{1 - KC_0}$$

with $\varrho_i = 0.26$; dashed line, $\varrho_i = 0$; where C^*, is the maximum surface coverage, ϱ_i is the irreversibly adsorbed fraction of the total adsorbed layer, C_0 is the bulk protein concentration, and K is the Langmuir equilibrium constant for the reversibly adsorbed protein.[21]

Thompson and Axelrod[29] used TIRF coupled with fluorescence correlation spectroscopy to investigate the adsorbed states of antibodies to dinitrophenyl (DNP) on fused silica coated with DNP conjugated BSA. The observed fluorescence spontaneously fluctuates with time as fluorescent molecules exit and enter a small, well-defined surface area. These fluorescence fluctuations can be autocorrelated. The shape of the fluorescence autocorrelation function depends on the bulk and surface diffusion coefficients, the adsorption and desorption rates, and the geometry of the observed area.[27] The binding of anti-DNP specifically to DNP on the surface was observed along with significant amounts of nonspecific binding. As pointed out by the authors, this is an interesting result that may also apply to binding between soluble proteins (antibodies or hormones, for example) and specific cell surface receptors. Other TIRF studies of antibody binding to antigen coated surfaces, however, have reported negligible amounts of nonspecific binding.[38]

The two examples discussed above illustrate the usefulness of the TIRF technique in characterizing multiply sorbed states often exhibited by adsorbing proteins. These examples also serve to illustrate the complexity of the protein adsorption process. The consistent findings of multiply sorbed states, including irreversibly adsorbed molecules, preclude the existence of a simple, dynamic equilibrium between adsorbed and desorbed states. It is also likely, that due to the long times required for adsorption or desorption to reach equilibrium, many reported isotherms do not reflect equilibrium conditions.

2. Protein Adsorption Kinetics

One of the primary advantages of using TIRF to examine protein adsorption is that continuous measurements can be made *in situ* and in real time so that adsorption and desorption rates can be unambiguously determined. Rates of protein adsorption can be governed by transport of protein to the surface as well as by adsorption kinetics. Thus, data must be properly analyzed to determine the extent to which mass transport limitations play a significant role. A detailed experimental and analytical study of the various transport effects observed in TIRF protein adsorption experiments is given by Lok et al.[21,22] using the transport analysis presented by Watkins.[36] Lok et al.[22] found the adsorption of BSA and fibrinogen from flowing solutions onto cross-linked PDMS to be diffusion limited for wall shear rates up to 420 sec^{-1} and solution protein concentrations up to 1000 mg%. This indicates that adsorption rates are so rapid that intrinsic kinetics cannot be observed; only diffusion rates from solution to the surface are observed. Subsequent studies by Cheng[37] have shown that adsorption from 1 mg% BSA solutions onto cross-linked PDMS is diffusion limited up to wall shear rates of 4000 sec^{-1}. Cheng also investigated the adsorption kinetics of BSA on a variety of other surfaces exhibiting various degrees of hydrophobic, electrostatic, and hydrogen-bonding capabilities. A summary of her results is given in Table 1. The findings of these investigations, as well as the results of other TIRF studies,[29] indicate that the initial rates of protein adsorption can often be predicted using only mass transfer considerations.

Extrinsic TIRF permits one to examine solutions containing mixtures of various proteins. Lok et al.[21,22] and Beissinger and Leonard[23,24] have demonstrated the use of TIRF to investigate the kinetics of adsorption from multiple protein solutions.

3. Protein Surface Diffusion

Burghardt and Axelrod[28] used TIRF coupled with photobleaching recovery to measure the surface diffusion of BSA adsorbed on quartz surfaces. A low-intensity beam was used to monitor adsorption until equilibrium was achieved, then a flash of a high-intensity beam was used to photobleach the fluorophores within the adsorbed layer.

Table 1

ADSORPTION CHARACTERISTICS OF BSA[37]

Surface		BSA Adsorption
Polydimethylsiloxane	CH_3 | $-Si-O-$ | CH_3	Diffusion limited up to wall shear rate of 4000 s^{-1}
Polydiphenylsiloxane	ϕ | $-Si-O-$ | ϕ	Diffusion limited up to wall shear rate of 4000 s^{-1}
Polycyanopropylmethylsiloxane	CH_3 | $-Si-O-$ | $(CH_2)_3$ | C ||| N	Diffusion limited up to wall shear rate of 4000 s^{-1}
Polymethylmethacrylate	CH_3 ... $C=O$ ... $O-CH_3$	Diffusion limited up to wall shear rate of 3000 s^{-1}
Polystyrenesulfonate	(phenyl ring with SO_3^-)	Kinetics-controlled for wall shear rate >300 s^{-1}
Polyethyleneoxide	$-CH_2-CH_2-O-$	No adsorption

The low-intensity beam was subsequently used to measure the fluorescence recovery. The fluorescence recovery is a measure of adsorption/desorption and surface diffusion rates. Burghardt and Axelrod found that their recovery curves could be approximated by a model that included irreversible, slowly reversible, and rapidly reversible components within the adsorbed layer. Using two different beam widths to examine the relative importance of surface diffusion and exchange, they found that the slowly reversible component was independent of beam width while the rapidly reversible component displayed a faster recovery in the narrow-beam experiments. This indicated that lateral surface diffusion contributes to the fluorescence recovery. The surface diffusion coefficient was determined to be approximately 5×10^{-9} cm²/sec. To date, this is the only technique used to measure the surface diffusion of adsorbed proteins.

B. Adsorbed Protein Conformation

As mentioned previously, in principle, TIRF can be coupled with a number of fluorescent techniques that have been used to study macromolecular conformations in solution. Work of this nature has only recently appeared in the literature and is discussed further on.

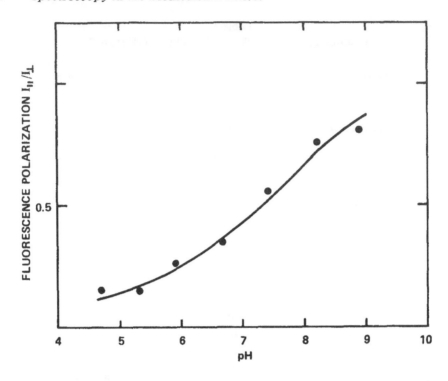

FIGURE 4.   Fluorescence polarization of fluorescein-labeled BSA adsorbed on PDMS as a function of pH.[37]

1. Fluorescence Polarization

Cheng[37] has presented preliminary results of polarization anisotropy studies investigating the relationship between the bulk solution pH and the refractive index of a layer of BSA adsorbed on PDMS. The presence of an adsorbed protein layer at the reflecting layer/bulk solution interface affects the intensity of the evanescent field. As discussed by Harrick,[7] the parallel-polarized component of the evansecent wave is sensitive to the refractive index of the adsorbed layer, whereas the electric field intensity of the perpendicularly polarized component is governed only by the properties of the two bulk media. Thus, if $l_{\parallel}$ and $l_{\perp}$ are the intensities of fluorescence emitted by the adsorbed layer when excited by parallel and perpendicular polarized light, respectively, then $l_{\parallel}/l_{\perp}$ depends on the refractive index of the adsorbed layer. Figure 4 shows the observed dependence of $l_{\parallel}/l_{\perp}$ on bulk-solution pH for a layer of BSA adsorbed on PDMS. The results are consistent with a decrease in adsorbed layer refractive index as pH increases. This would result if the adsorbed layer was expanding with increasing pH. The isoelectric point of BSA is 4.7. As pH increases above this value, the BSA molecule acquires an increasingly negative charge. Thus, forces arising from electrostatic repulsion could be expanding the adsorbed layer.

Burghardt[25] has measured fluorescence depolarization subsequent to polarized excitation to examine the rotational diffusion of chromophores attached to BSA both in solution and adsorbed on a quartz surface. The fluorescence depolarization technique has been used to study protein rotational diffusion and the internal flexibility of macromolecules.[39-42] The decay of the emitted fluorescence polarization anisotropy with time following polarized excitation is due to rotational diffusion of the chromophore. The time decay can be related to a rotational diffusion coefficient. Using fluorescence polarization measurements together with the known rotational diffusion coefficient of bulk-dissolved BSA, the local rotational diffusion coefficients were calculated for chromophores attached to BSA in solution and BSA adsorbed on a quartz surface. The

Table 2
ROTATIONAL DIFFUSION COEFFICIENTS OF
FLUOROPHORES BOUND TO SOLUTION-DISSOLVED AND
SURFACE-ABSORBED BSA[25]

	$D_{rotational} \times 10^{-9}$ (sec^{-1})	
Fluorophore	Bound to solution-dissolved BSA	Bound to surface-adsorbed BSA
Dansyl	0.0065	0.007
4-Chloro-7-nitrobenzo-2-oxadiazol	0.075	0.013
Rhodamine	0.115	0.07
Eosin	1.3	0.10

results for four different fluorescent labels are presented in Table 2. The BSA used for this study was nonspecifically labeled with an average of 5 to 10 labels per molecule. The rotational diffusion coefficients vary widely for the different fluorophores used. All the labels except dansyl showed a significant reduction in rotational mobility upon adsorption. It is not clear why dansyl behaves differently. The restricted rotational motion of the probes is attributed to steric hindrance either from the surface or from conformational changes in the adsorbed molecules.

2. Fluorescence Energy Transfer

Fluorescence energy transfer is a well-known fluorescence technique useful for estimating distances between chromophores up to separations of approximately 60 Å.[43] When the emission spectra of one fluorophore (the donor) overlaps with the absorption spectra of another (the acceptor), the emission of the excited donor will result in the excitation and fluorescence of the acceptor. By measuring the decrease in fluorescence of the donor or the increase in fluorescence of the acceptor, the energy-transfer efficiency can be determined. The efficiency is highly sensitive to the distance separating the two fluorophores. Burghardt and Axelrod[26] have applied the use of this technique in the study of doubly labeled BSA adsorbed on quartz. Two donor-acceptor pairs were used — 4-chloro-7-nitro-2,1,3-benzoxadiazole/rhodamine and dansyl/eosin. The BSA was nonspecifically labeled with several donors and acceptors per BSA molecule. The authors observed decreased energy-transfer efficiencies following adsorption and concluded that the decrease was due possibly to conformational changes between the bulk-dissolved BSA and the surface-adsorbed BSA. The authors acknowledge, however, that several assumptions were made in arriving at the above conclusion. An accurate calibration of the relationship between the fluorescence intensity of acceptors attached to surface-adsorbed BSA and the absolute surface-adsorbed BSA concentration is also required. A calibration was performed but it may be faulty since it was conducted with an adsorbed BSA layer in contact with pure buffer while the energy-transfer experiments were performed with an adsorbed BSA layer in equilibrium with BSA in the bulk solution. The quantum yield of fluorophores attached to surface-adsorbed BSA has been observed to depend significantly on the bulk-solution BSA content in other TIRF experiments.[21] Nevertheless, the feasibility of coupling TIRF with the fluorescence energy-transfer technique has been demonstrated. As with the fluorescence-polarization technique discussed earlier, the energy-transfer technique should prove most useful in cases where specifically labeled proteins, with one donor and one acceptor in identical positions on each protein molecule, can be used. It may then be possible to make more detailed statements concerning conformational changes between bulk and surface-adsorbed molecules.

IV. CONCLUSIONS

In the future, TIRF will certainly be applied to further studies on the equilibrium, kinetic, and surface diffusion characteristics of macromolecules adsorbing to solid surfaces. As a logical outgrowth of these studies, TIRF is just now finding applications in investigations of adsorbed protein conformation. Fluorescence techniques traditionally used to examine solution protein conformations can be coupled with TIRF to study the conformational aspects of surface-adsorbed proteins. The use of TIRF in fluorescence-polarization and energy-transfer studies has been demonstrated.

The interaction of specific antibodies with protein antigens is another technique that has been applied to the study of the dynamic and steady-state aspects of solution protein conformations.[44] It may be of interest to investigate the effect of antigen protein adsorption on the interactions of specific antibodies in solution. The sensitivity of TIRF would allow the detection of even subtle changes in antibody affinities for the adsorbed protein.

TIRF may also find future applications in the study of the interactions of cell surfaces with other surfaces, either solid substrates or other membrane surfaces. Axelrod[45] has demonstrated a TIRF microscopy technique to visualize specifically labeled molecules in regions of contact between cultured cells and glass substrates while avoiding the fluorescence of the cell interior and extracellular debris. This may prove useful, for instance, in the study of observed cytoskeletal rearrangements following adsorption of a cell to a solid substrate.[46] Weis et al.[47] have used TIRF microscopy to examine the regions of contact between rat basophil leukemia cells and lipid monolayers coated on the surfaces of quartz slides. When lipid haptens were included in the monolayers, serotonin release was triggered from the cells in the presence of specific anti-hapten IgE antibodies. The regions of contact were visualized by fluorescently labeling the IgE antibodies. Such uses of TIRF may find applications in immunological studies of receptor-mediated cell triggering.

Although no single technique can provide all the desired information concerning the interactions of biological macromolecules with solid surfaces, TIRF can provide information that other commonly used techniques cannot. The intent of this chapter is to enumerate some of the applications where the advantages of using TIRF are best realized.

REFERENCES

1. Baier, R. E. and Dutton, R. C., Initial events in interactions of blood with a foreign surface, *J. Biomed. Mater. Res.*, 3, 191, 1969.
2. Packham, M. A., Evans, G., Glynn, M. F., and Mustard, J. F., The effect of plasma proteins on the interaction of platelets with glass surfaces, *J. Lab. Clin. Med.*, 73, 686, 1969.
3. Stoner, G. E., Srinivasan, S., and Gileadi, E., Adsorption inhibition as a mechanism for the antithrombogenic activity of some drugs. I. Competitive adsorption of fibrinogen and heparin on mica, *J. Phys. Chem.*, 75, 2107, 1971.
4. Bock, H. G., Skene, P., Fleischer, S., Cassidy, P., and Harshman, S., Protein purification: adsorption chromatography on controlled pore glass with the use of chaotropic buffers, *Science*, 191, 380, 1976.
5. Askenase, P. W. and Leonard, E. J., Solid phase radioimmunoassay of human β 1C globulin, *Immunochemistry*, 7, 29, 1970.
6. Adam, G. and Delbruck, M., Reduction of dimensionality in biological diffusion processes, in *Structural Chemistry and Molecular Biology*, Rich, A. and Davidson, N., Eds., W. H. Freeman, San Francisco, 1968, 198.
7. Harrick, N. J., *Internal Reflection Spectroscopy*, John Wiley & Sons, New York, 1967.

8. Hirschfeld, T., Total reflection fluorescence (TRF), *Can. Spectrosc.*, 10, 128, 1965.
9. Harrick, N. J. and Loeb, G. I., Multiple reflection fluorescence spectrometry, *Anal. Chem.*, 45, 687, 1973.
10. Brash, J. L. and Lyman, D. J., Adsorption of plasma proteins in solution to uncharged, hydrophobic polymer surfaces, *J. Biomed Mater. Res.*, 3, 175, 1969.
11. Vroman, L., Adams, A. L., Klings, M., and Fischer, G., Fibrinogen, globulins, albumin and plasma at interfaces, *Adv. Chem. Ser.*, 145, 255, 1975.
12. Fair, B. D. and Jamieson, A. M., Studies of protein adsorption on polystyrene latex surfaces, *J. Colloid Interface Sci.*, 77, 525, 1980.
13. Dillman, W. J., Jr. and Miller, I. F., On the adsorption of serum proteins on polymer membrane surfaces, *J. Colloid Interface Sci.*, 44, 221, 1973.
14. Brash, J. L., Uniyal, S., and Samak, Q., Exchange of albumin adsorbed on polymer surfaces, *Trans. Am. Soc. Artif. Intern. Organs*, 20, 69, 1974.
15. Brash, J. L. and Davidson, V. J., Adsorption on glass and polyethylene from solutions of fibrinogen and albumin, *Thromb. Res.*, 9, 249, 1976.
16. Weathersby, P. K., Horbett, T. A., and Hoffman, A. S., Fibrinogen adsorption to surfaces of varying hydrophilicity, *J. Bioeng.*, 1, 395, 1977.
17. Lee, R. G., Adamson, C., and Kim, S. W., Competitive adsorption of plasma proteins onto polymer surfaces, *Thromb. Res.*, 4, 485, 1974.
18. Ihlenfeld, J. V. and Cooper, S. L., Transient *in vivo* protein adsorption onto polymeric biomaterials, *J. Biomed. Mater. Res.*, 13, 577, 1979.
19. Walton, A. G. and Kotisko, B., Protein structure and the kinetics of interaction with surfaces, *Adv. Chem. Ser.*, 199, 245, 1982.
20. Gendreau, R. M., Leninger, R. I., Winters, S., and Jakobsen, R. J., Fourier transform infrared spectroscopy for protein-surface studies, *Adv. Chem. Ser.*, 199, 371, 1982.
21. Lok, B. K., Cheng, Y., and Robertson, C. R., Total internal reflection fluorescence: a technique for examining interactions of macromolecules with solid surfaces, *J. Colloid Interface Sci.*, 91, 87, 1983.
22. Lok, B. K., Cheng, Y., and Robertson, C. R., Protein adsorption on crosslinked polydimethylsiloxane using total internal reflection fluroescence, *J. Colloid Interface Sci.*, 91, 104, 1983.
23. Beissinger, R. L. and Leonard, E. F., Plasma protein adsorption and desorption rates on quartz: approach to multi-component systems, *Trans. Am. Soc. Artif. Intern. Organs*, 27, 225, 1981.
24. Beissinger, R. L. and Leonard, E. F., Sorption kinetics of binary protein solutions: general approach to multicomponent systems, *J. Colloid Interface Sci.*, 85, 521, 1982.
25. Burghardt, T. P., Fluorescence depolarization by anisotropic rotational diffusion of a luminophore and its carrier molecule, *J. Chem. Phys.*, 78, 5913, 1983.
26. Burghardt, T. P. and Axelrod, D., Total internal reflection fluorescence study of energy transfer in surface-adsorbed and dissolved bovine serum albumin, *Biochemistry*, 22, 979, 1983.
27. Thompson, N. L., Burghardt, T. P., and Axelrod, D., Measuring surface dynamics of biomolecules by total internal reflection fluorescence with photobleaching recovery or correlation spectroscopy, *Biophys. J.*, 33, 435, 1981.
28. Burghardt, T. P. and Axelrod, D., Total internal reflection/fluorescence photobleaching recovery study of serum albumin adsorption dynamics, *Biophys. J.*, 33, 455, 1981.
29. Thompson, N. L. and Axelrod, D., Immunoglobulin surface binding kinetics studied by total internal reflection with fluorescence correlation spectroscopy, *Biophys. J.*, 43, 103, 1983.
30. Van Wagener, R. A. and Andrade, J. D., Total internal reflection fluorescence (TIRF) intrinsic spectroscopy in the UV by excitation of tyrosine and tryptophan in proteins, in *Surface and Interfacial Aspects of Biomedical Polymers*, Andrade, J. D., Ed., Plenum Press, New York, in press, 1985.
31. Iwamoto, G. K., Van Wagenen, R. A., and Andrade, J. D., Insulin adsorption: intrinsic interfacial fluorescence, *J. Colloid Interface Sci.*, 86, 581, 1982.
32. Van Wagenen, R. A., Rockhold, S., and Andrade, J. D., Probing protein adsorption. Total internal reflection intrinsic fluorescence, *Adv. Chem. Ser.*, 199, 351, 1982.
33. Beissinger, R. L. and Leonard, E. F., Immunoglobulin sorption and desorption rates on quartz: evidence for multiple sorbed states, *J. Am. Soc. Artif. Intern. Organs*, 3, 160, 1980.
34. Watkins, R. W. and Robertson, C. R., A total internal reflection fluorescence technique for the examination of protein adsorption, *J. Biomed. Mater. Res.*, 11, 915, 1977.
35. Cheng, Y., Lok, B. K., and Robertson, C. R., Interactions of macromolecules with surfaces in shear fields using visible wavelength total internal reflection fluorescence, in *Surface and Interfacial Aspects of Biomedical Polymers*, Andrade J. D., Ed., Plenum Press, New York, in press, 1985.
36. Watkins, R. W., A Total Internal Reflection Fluorescence Technique for the Study of Protein Adsorption, Ph.D. thesis, Stanford University, Stanford, Calif., 1976.
37. Cheng, Y., A Total Internal Reflection Fluorescence Study of Bovine Serum Albumin Adsorption onto Polymer Surfaces, Ph.D. thesis, Stanford University, Stanford, Calif., 1983.

38. Kronick, M., Sensing of Structure-Specific Binding onto Functionalized Quartz Surfaces Using Total Internal Reflection Fluorescence, Ph.D. thesis, Stanford University, Stanford, Calif., 1974.
39. Munro, I., Pecht I., and Stryer, L., Subnanosecond motions of tryptophan residues in proteins, *Proc. Natl. Acad. Sci. U.S.A.,* 76, 56, 1979.
40. Lovejoy, C., Holowka, D. A., and Cathou, R. E., Nanosecond fluorescence spectroscopy of pyrene-butyrate-anti-pyrene antibody complexes, *Biochemistry,* 16, 3668, 1977.
41. Mendelson, R. A., Morales, M. F., and Botts, J., Segmental flexibility of the S-1 moiety of myosin, *Biochemistry,* 12, 2250, 1973.
42. Yguerabide, J., Epstein, H. F., and Stryer, L., Segmental flexibility in an antibody molecule, *J. Mol. Biol.,* 51, 573, 1970.
43. Stryer, L., Fluorescence energy transfer as a spectroscopic ruler, *Ann. Rev. Biochem.,* 47, 189, 1978.
44. Hurrell, J. G. R., Smith, J. A., and Leach, S. J., Immunological measurements of conformational motility in regions of the myoglobin molecule, *Biochemistry,* 16, 175, 1977.
45. Axelrod, D., Cell-substrate contacts illuminated by total internal reflection fluorescence, *J. Cell Biol.,* 89, 141, 1981.
46. Rees, D. A., Lloyd, C. W., and Thom, D., Control of grip and stick in cell adhesion through lateral relationships of membrane glycoproteins, *Nature (London),* 267, 124, 1977.
47. Weis, R. M., Balakrishnan, K., Smith, B. A., and McConnell, H. M., Stimulation of fluorescence in a small contact region between rat basophil leukemia cells and planar lipid membrane targets by coherent evanescent radiation, *J. Biol. Chem.,* 257, 6440, 1982.

Chapter 9

APPLICATIONS OF FLUORESCENCE TECHNIQUES TO AUTOMATED BIOLOGICAL ANALYSIS

Joseph L. Giegel

TABLE OF CONTENTS

I. INTRODUCTION

In contrast to the previous chapter, this final chapter discusses an area of spectroscopy (flourescence coupled with enzyme-linked immunoassay techniques) which is finding widespread use in the medical diagnostic industry.

Although fluorescence techniques are inherently more sensitive than colorimetric methods of analyses, they have not achieved widespread acceptance in the clinical laboratory. This is due in part to contamination problems that were frequently encountered in many of the early methods used by the clinical laboratory. Some of these problems can be avoided through the use of automation and disposable pipettes and tubes for analyses. Interest has been expressed in the kinetic measurements of enzyme activity using fluorescence analyses. However, this also has not achieved wide popularity, principally because the increased sensitivity of fluorescence is not required. Ultraviolet (UV) measurements have proven to be quite adequate and, in addition, UV absorbance is not influenced by changes in temperature as in fluorescence.

Many of the analytes of clinical interest can be determined by radioimmunoassay (RIA), and this became the dominant method for analysis of compounds with low concentrations in clinical samples. Although the growth of RIA has been quite extensive, alternative techniques have been sought, principally because of the complexities of radioactive waste and the short shelf life of reagents. Fluorescence analysis is, therefore, undergoing a renaissance in the laboratory with two principle approaches. The first is the replacement of the radioisotopic label with a fluorescent label and performing similar types of analyses as RIA. This has been well accepted, particularly for drug analysis. A second approach has been the use of enzymes in place of radioactive tags. A high degree of sensitivity can be achieved when these enzyme labels are assayed by fluorometric techniques. Sensitivity at least as good as that obtained by RIA has now been achieved with enzyme-labeled systems. This chapter reviews one such system which uses enzyme labels and fluorescence detection to achieve very rapid and highly sensitive immunoassays. The system has been termed "Radial Partition Immunoassay" since the entire assay is conducted on a small segment of filter paper and uses radial chromatography to conduct an immunoassay.[1]

II. SYSTEM DESCRIPTION

A schematic outline of the immunoassay procedure is shown in Figure 1. The first step of each procedure is to coat antibody onto a segment of glass-fiber filter paper. Usually a central zone of the paper is coated; however, the entire paper may be coated as well. After coating, the filter paper is dried and the antibody is fixed to the glass fibers. As illustrated in Figure 1, three different types of assays can be conducted using these filter paper tabs. In a *sequential* assay, sample is first applied to the dry tab. During this period, antigen in the sample can react with the antibody. Following this step, enzyme-labeled antigen is added to fill the remaining antibody sites. The third step is the addition of the substrate wash fluid for the enzyme to the center area of the tab. The addition of the substrate wash accomplishes two functions. The first is to radially chromatograph any unbound enzyme-labeled antigen to the edge of the paper and the second is to initiate the enzyme reaction leading to the development of fluorescence in the center of the tab. This fluorescence is read over a small section of the center area and the rate of change of fluorescence is proportional to the concentration of enzyme-labeled antigen on the tab.

A second type of assay is conducted in a *competitive* mode. In this case the sample containing antigen and enzyme-labeled antigen are premixed prior to addition to the filter-paper tab. The labeled and unlabeled antigen then compete for antibody sites.

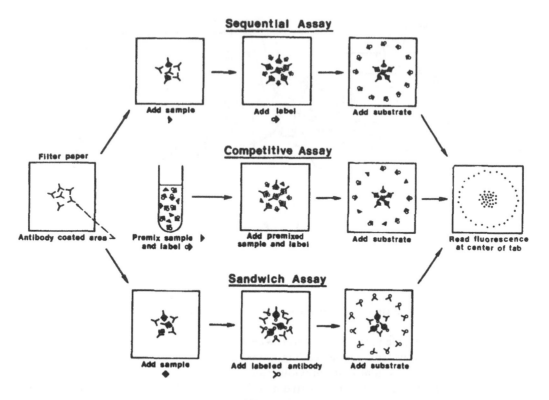

FIGURE 1. Outline of radial partition immunoassay.

After a brief interval, substrate is again added and the fluorescence is quantitated as for the sequential assays. A third type of assay is termed a *sandwich* assay in which two antibodies are used. In addition to the antibody coated onto the filter paper, a second antibody is labeled with an enzyme. In this type of assay, the first step is similar to the sequential assay in that sample is added to the dry filter-paper tab. To conduct a sandwich assay, it is necessary to have antigens which have more than one binding site. After the first step, enzyme-labeled antibody is then added which binds to antigen bound in the first step. The third step, addition of substrate, is common to the previous assays.

Each of the three types of assays described above has been found to produce very rapid immunoassays. Typically, each assay can be completed in less than 10 min. Sequential assays are very useful for low concentration analytes such as the cardiac drug digoxin in serum. Competitive assays are useful for drugs present at higher concentrations or where it is necessary to strip the antigen from the serum-binding proteins prior to immunoassay; for example, the measurement of thyroxine in serum. The third type of assay is very useful for large multivalent antigens; for example, the analysis of ferritin in serum.

III. INSTRUMENTATION

A fully automated instrument that can conduct each of these assays is commercially available under the trade name Stratus® and a schematic outline is shown in Figure 2. Sample, enzyme-labeled conjugate, and substrate addition are all handled by the instrument. In addition, there is a read station which takes readings over a 20-sec interval, computes the rate of change of fluorescence, and converts this into concentration of the antigen. One of the key features of the instrument is its stability, achieved by

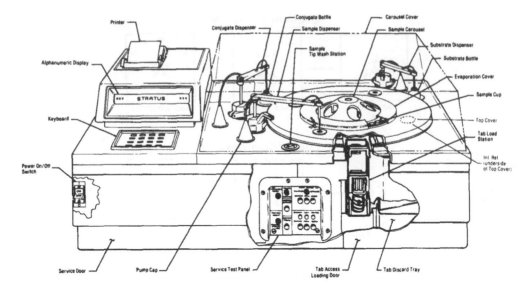

FIGURE 2.   Description of Stratus® instrument.

the inclusion of a stable reference material which is fluorescent. Prior to the beginning of each run on the instrument, a reading is taken on this reference material and any change in gain in the system from previous runs is corrected for.

The optical part of the Stratus® instrument is quite simple, as shown in Figure 3. A low-pressure mercury lamp produces light at 366 nm as the primary wave length. This is transmitted to the tab by a dichroic mirror and the fluorescence which peaks at 470 nm passes through the dichroic mirror into a photomultiplier tube. The efficiency of this system could be expanded in the future to further improve the sensitivity of the system, should this be required. In addition, it would be conceivable to use some other techniques such as photon counting to further improve sensitivity.

IV. RESULTS

A typical rate tracing of a calibration curve for the drug digoxin is shown in Figure 4. These assays are conducted in a sequential mode. The highest rates are observed at the lowest concentrations of the drug which is typical of this type of immunoassay. As the drug concentration increases, less of the enzyme-labeled antigen is able to bind to the filter paper and, therefore, the rates decrease. Rates have been observed to be fairly linear over relatively long periods of time (> 5 min). However, the typical assay is conducted by reading a 20-sec segment of each of these curves. A very good signal-to-noise ratio is observed with the system permitting short time intervals of rate measurement.

An interesting aspect of the immunoassay system is the low nonspecific binding observed. The glass-fiber filter paper is fairly inert and once coated with antibody becomes even more inert. Figure 4 shows the blank rate with such tabs. In this case, the enzyme-labeled digoxin is added to a tab which does not contain specific antibody. This low blank rate observed illustrates the very low nonspecific binding in the system.

The data shown in Figure 4 are graphed in a more conventional manner in Figure 5 and compared to a curve obtained with a radioisotopic label. In this case, both the Stratus® and radioimmunoassay use the same antibody. Stratus® uses enzyme-labeled digoxin, and RIA uses ^{125}I-labeled digoxin. Similar curves can be generated with both systems and the sensitivity of both is essentially identical. In principle, each la-

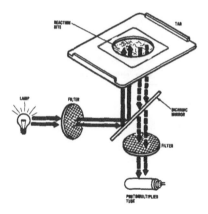

FIGURE 3.　Schematic of Stratus® optical system.

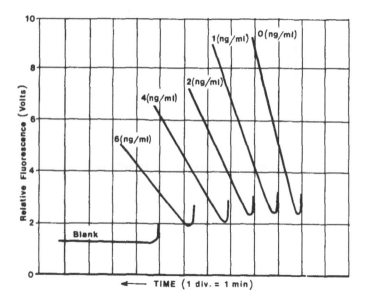

FIGURE 4.　Fluorescent rate curves for a digoxin assay.

beled molecule can be visualized in the enzyme immunoassay system as opposed to the limited number of radioactive decays in the RIA system. Because of this, the sensitivity of enzyme-labeled systems can inherently be greater than radioisotopic-labeled systems.

One of the problems which has limited the acceptance of fluorescence techniques in the clinical laboratory has been the negative experiences many clinical chemists encountered with contamination in the early use of fluorescence techniques. The system described is able to eliminate such contamination problems in several ways, principally by using kinetic methods of analysis. Extensive method comparisons have been conducted against RIA methods to assess potential interferences. Figure 6 shows an example of a digoxin comparison between the Stratus® and an RIA system. This comparison is quite good. Additional comparisons have been conducted with other types of assays and none of the contamination problems experienced in the past have been apparent.

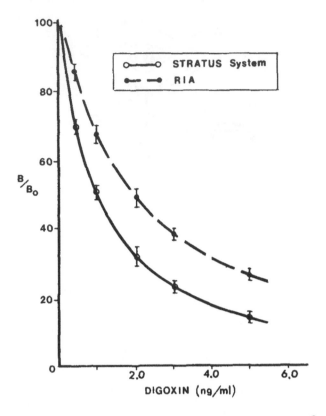

FIGURE 5.   Calibration curves for an RIA assay and a Stratus®
assay.

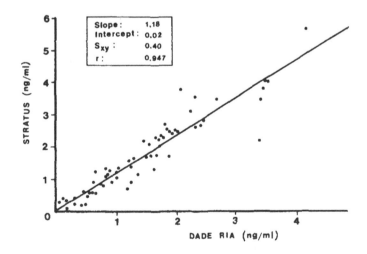

FIGURE 6.   Correlation between on RIA procedure and the Stratus®
procedure for digoxin in serum.

V. LARGE MOLECULE IMMUNOASSAYS

A particularly exciting area for the application of the sensitivity of radial partition
immunoassay is in the analysis of antigens with high molecular weight. Typically these
antigens have multiple antibody-binding sites and can be done using sandwich assays.
In sandwich assays, the signal starts low and increases with increasing concentration

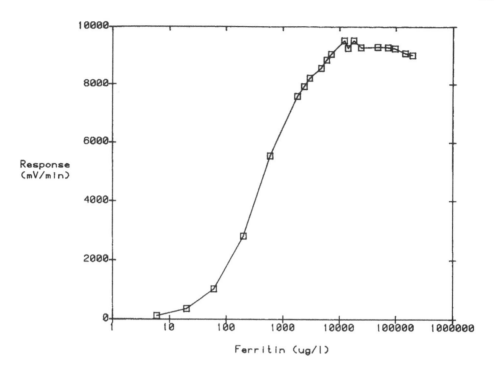

FIGURE 7. Calibration curve for a ferritin sandwich assay.

of antigen, as opposed to the sequential or competitive assays in which the signal starts high and decreases. In many fluorescence techniques sensitivity can be limited by non-specific background fluorescence. Because of the nature of the glass-fiber filter paper used, this background signal can be made very low. An example of a calibration curve for ferritin is shown in Figure 7. Ferritin has a molecular weight of approximately 450,000 and is present in clinical samples in the nanogram-per-milliliter concentration range. This curve shows the inherent sensitivity of the sandwich assay as well as the large dynamic range which can be achieved. Sensitivity has been calculated on repeated samples to be less than 1 ng/mℓ.

One of the traditional problems with sandwich assays has been what has been termed the "high-dose hook effect"[2] in which the response actually decreases as the concentration increases above a certain point. Figure 7 illustrates that even at very high concentrations of antigen, there is only slight and insignificant decrease in signal, thus, high sensitivity can be combined with a large dynamic range.

The future of fluorescent techniques in biological analysis is extremely promising. Most of the advances will combine immunology with fluorescence to achieve the specificity inherent in immunological techniques and the sensitivity which can be achieved with fluorescence techniques. Further improvements in sensitivity can be expected as increased use of photon-counting techniques and time resolved fluorescence become widely accepted.

REFERENCES

1. Giegel, J. L. et al., Radial partition immunoassay, *Clin. Chem.*, 28, 1894, 1982.
2. Miles, L.E.M. et al., Measurement of serum ferritin by a 2-site immunometric assay, *Anal. Biochem.*, 61, 209, 1974.

INDEX

Printed and bound by CPI Group (UK) Ltd, Croydon, CR0 4YY
22/10/2024
01777630-0003